SOUND RECORDING

Sound Recording
FROM MICROPHONE TO MASTER TAPE

DAVID TOMBS

David & Charles

Newton Abbot London North Pomfret (Vt)

ISBN 0 7153 7954 2

British Library Cataloguing in Publication Data

Tombs, David
   Sound recording.
   1. Sound – Recording and reproducing
   I. Title
   621.389'32   TK7881.4

   ISBN 0–7153–7954–2

Printed in Great Britain by
REDWOOD BURN LIMITED
Trowbridge & Esher
for David & Charles (Publishers) Limited
Brunel House Newton Abbot Devon

Published in the United States of America
by David & Charles Inc
North Pomfret Vermont 05053 USA

Contents

CONTENTS

CONTENTS

The equipment
Processing to masters
Dub-editing
Cut-editing

Introduction—A Super Sound

'What a beautiful sound, and what a super recording!'—words which every sound recordist likes to hear, especially when they refer to his own handiwork. Getting good sounds on to tape in the best way possible is the consuming passion of the recordist and, equipped with the right gear and knowledge, sound recording can be an enjoyable and highly rewarding activity.

There are lots of things we can do with tape recorders (apart from using them as bookends and paperweights). We can record items from the radio (for replay some time later), or we can record original sounds—a visiting choir or orchestra, the stage play at the Community Centre and the talk given at the topping-out or foundation-stone ceremony, or the special drama recorded for the local radio station. There are the many and varied sounds of nature—water, wind and wildlife.

Tape machines can be used for echo, too. When a three-head machine is used for recording, the results are available for replay (monitoring) almost instantaneously. There is a small time delay, governed by the tape speed and the distance between the recording and replay heads, and this can be used to produce a multiplicity of interesting echo effects.

Another major use for your tape machine is editing—the rearrangement and cutting of your recordings—and when your friends see how easy it is to edit tape, they too will probably want to have a go! I am still finding new and interesting techniques with tape, and I have been recording and editing for more than 15 years. It can be very satisfying when, after getting to work on an interview tape which seems at first to be an incoherent babble in an illogical sequence, there emerges an easy-flowing and concisely told story or interview. There are the amusing moments when, having made a tape loop of sound effects, one discovers that even stranger effects arc produced if the tape is played backwards. Amazing sounds are produced by playing a recording of a simple sound at one-eighth its normal speed, or by editing together recording tape and blank spacer in a quickly alternating fashion.

Gimmicks aside, there is the pleasure of listening to one's own first-

9

class recordings of a wealth of sounds, from the scratchings of a wood-boring beetle to the glorious sound of a full orchestra. For me, the most pleasurable sounds remain those of nature.

Whatever field one is interested in, the final recorded result can be no better than the signals delivered to the recording machine, and their quality depends solely on the skill of the recordist and on his choice and use of microphones. There is no reason why the novice recordist should not produce first-class recordings; in fact, some recordings by novices are sufficiently outstanding to be broadcast on national radio. Good sound recordists strive to produce recordings which are limited only by the recording medium itself, or by environmental factors over which one has no control. There are now many practised amateurs producing work of a quality much admired in professional circles. In the professional sphere time is always at a premium. The amateur is not encumbered by time-costing—only by the limits of his patience.

Good tape recorders alone do not necessarily mean good recordings. It is in signal-gathering, the amplifying and the processing of sound signals that good recordings are made. The recordist can buy, or construct, a number of accessories which make a world of difference to the quality of recordings and to the versatility of recording machines, extending, particularly in the case of cassette machines, the range of recording activities that can be undertaken. Electronics have advanced to a stage where even non-technically minded enthusiasts can easily build various pieces of equipment from the raw component parts or from pre-assembled and tested units. Power amplifiers, pre-amplifiers and a host of other units are available off-the-peg.

When I first became interested in electronics and hi-fi things were very different and altogether more difficult. My first hi-fi amplifier, a 20W valve job using push-pull 6L6s operating on 350V dc, was a home-built affair which gave off enough heat to warm the garden shed on a winter's day. Its construction involved hours of drilling and the cutting of large holes in an aluminium chassis for the many valves, capacitors and other bits and pieces. The output transformer was as large as the mains transformer and both were heavy enough, were they to have been accidentally dropped, to have gone clean through the tool drawer and give any woodworm in the floor an appreciable headache! I was sixteen years old when I built this amazing piece of equipment, and the hinged, lift-up workbench bears to this day the scars from those happy hours of construction. I had a 78rpm turntable with a magnetic pick-up, a moving-coil microphone mounted in a tin box with a perforated front, and two 254mm loudspeakers mounted in 914 × 914mm open-back cabinets. With this equipment and a pile of 78rpm records I provided 'music and mike' at dances, weddings, garden parties and fêtes. But I could not *record*.

Today, with far less weighty equipment (thanks to transistors) and for a modest outlay, it is quite within the capabilities of most keen amateurs to enjoy such activities and also to make their own recordings of music, voice and natural effects. The recorder need not be elaborate, for success depends not only on the complexity of the equipment but also on the competence of the user. In the sphere of photography, which has so many parallel features with sound recording, accomplished masters in the art of composition and light utilisation have achieved amazingly good pictures with pin-hole cameras. Similarly, the expert sound recordist aims for perfection, though satisfyingly good recordings are more likely to be the reward. Electronic video recording is making advances in the amateur motion-picture world, but sound recording scores markedly over photography in that the recorded results are replayable as soon as the tape is rewound. Also, recording tape is relatively cheap and, with care, it can be reused thousands of times. When eventually it does wear out it can be stretched like bootlace and used in the garden for tying up plants.

Many recorders have auto-level control, so that knob-twiddling is eliminated. Some people like knobs—the more there are the more they like them—and each knob on those vast control panels in sound studios has its purpose. We shall be turning a few of these knobs later on.

To make good recordings it is not absolutely necessary to know how a tape recorder works—no more necessary than it is to know how a car engine works to be a good driver. We shall not lift the lid on the tape recorder and dive into its mechanics or into its recording process— there are already a number of books on those subjects—though we shall concern ourselves with the performance of, and recognition of, faults on, recording machines. Our primary concern is with microphones and sound and the problems of getting it on to tape. However, the aspiring recordist should know how to put the tape on a machine, how to press the recording button and how to record at a level just below the distortion point on the tape.

There are, of course, vast differences between cheap cassette recorders and studio master recording machines—apart from the price. As with most equipment, when it comes to quality one gets what one pays for. When considering recording machines it usually pays in the long run to acquire equipment which offers a greater range of facilities than you think are necessary, because further along the road of experience the additional facilities may well prove to be indispensable. This is not to say that one should obtain a monster 64 track machine just in case it

11

may come in handy one day, when a good three-head stereo machine can provide adequate facilities for the foreseeable future.

Perhaps you already have a tape recorder: a portable, or one fixed in a 'music centre' or radiogram. It is easy to record items from the radio and from the record player in music centres, because only the appropriate buttons have to be pushed, while the hard graft has already been done by others. Many deceptively simple works of art are the result of months of effort by a number of people and, for this reason alone, it is only fair to respect rules on copyright. In the same way that certain products carry Government health warnings, commercial records carry the copyright warning, and radio-originated material must not be used other than for personal and family entertainment and for educational purposes. Anyway, there is no true achievement in cribbing someone else's work—a recording machine can be put to far more inventive uses!

Whether the recordist uses his machine as a personal diary for on-the-spot reports, for recording the conversation when he phones an official about the smell from the drains, or for recording grandma on holiday ordering dinner for the family at that little restaurant in France, he is obtaining original material—material which, like snapshots in the family album, can be drawn on as individual items for nostalgic reflection, or used with other recordings from the collection to produce programmes for family occasions. Broadcasting organisations assemble very interesting programmes from their archive material, and so too can the amateur.

The tape recorder is now small enough and light enough to carry about as one carries a camera. You could be in the market square in Main Street when, by chance, a wheel of a passing safari-park van falls off and rolls into the fruit stall; the van door bursts open, releasing a roaring lion into apples, oranges and market-day shoppers laden with baskets of eggs. Then, from your vantage-point on top of the public telephone-box, you record the havoc as shoppers less agile than yourself make for safety. Later, and after your recording has been heard on the radio news programme, you can, whenever you wish, relive the excitement of the incident in the market square.

Good sound reproduction is as important as good-quality recording and we shall be dealing with reproduction as well as sound recording. The first few chapters are technical. They explain sound: what it is and how we hear it. We find out about microphones and their directional responses, and how to use them in difficult or noisy locations. Manufacturers' data will be looked at to see how they can help us to choose a microphone. We investigate electrical noises and discover some solutions. Recording machines come under scrutiny, as do recording tapes, headphones and loudspeakers.

The later chapters deal with the more artistic and creative uses of microphones and recording systems. Stereo is analysed in depth, and the recordist following the detailed description of how to go about it should be able to make first-class stereo recordings. We take each of the main subjects—talks, drama, music and wildlife—in separate stages and discuss some recording techniques. Good windshielding techniques are also demonstrated. The final chapter, on 'The Home Studio', possibly the most interesting to the hi-fi recording enthusiast, deals with the acoustic treatment of the listening-room or studio, and the equipment needed. Processing our recordings is covered, together with tape-editing and dubbing.

The beginner should not find the text too difficult to follow, even if much of it is fairly technical and detailed. As well as giving a simple guide for the novice who wants to know how to make good recordings on his current equipment, how to improve his present standard and how to avoid some of the pitfalls, information of a more advanced nature is provided because, in the space of a few weeks, the novice can progress to the rank of amateur recording enthusiast—towards whom this work is primarily aimed. The more advanced details should extend the keen amateur and the tricks of the trade may help him to improve his technique. The professional sound recordist extending his horizons may also benefit from some of the tips included especially for him.

In our journey from microphone to master tape we shall encounter many problems and discover not a few solutions. For instance, did you know that the noise caused by a humming ventilation fan, in an area where you might wish to record, can be neutralised? And what about the wind and all those ruined recordings, the traffic noise on your outside interview tape and the foreign radio station which breaks through when you use a long microphone cable on your recording machine? What can you do about that noise you get when the performers are cavorting about on the stage? What sort of microphone should you use, and where should you place it, when you record a choir? How can you record the carnival procession in genuine stereo? If you would like to know how I would do it—read on.

1

The Nature of Sound

Air, the carrier

It would be difficult to record sound were it not for the vehicles which transport it to our microphones and to our ears. Sound is conducted along and through solids, liquids and gases. It will not travel in a vacuum.

Vibrations (and that is what sound-waves are) travel easily in liquids and solids, and recordists working in specialised fields take advantage of this. However, although a few wildlife recordists may dip their makeshift hydrophones (microphones adapted using certain rubber goods) into fish tanks and duck ponds, most of us are more familiar with the last in the group—gases. The gas around us is a mixture of nitrogen, oxygen, argon and water vapour; the ratios need not concern us. We call it air.

Air is not weightless: it weighs about $1.3 kg/m^3$ and this is sufficient, when it is on the move (wind), to cause our plastic macs to flap and to move other obstacles in its path. Moving air can also cause noise on recordings, and special shielding techniques have to be adopted to prevent it from blowing into microphones used outside. Air is very flexible and its weight is not generally referred to, but the result of its weight *is*: this is air pressure.

The pressure of the air around us—atmospheric pressure—varies from day to day, depending on weather conditions and altitude. At sea level the pressure is commonly 1 bar, or $1 kg/cm^2$. What this means in practical terms is that for every metre-square area there is a force of 10 tonnes; this force, in pressure differential terms, is very useful in preventing aircraft from falling out of the sky. What has this to do with sound recording? Well, without the pressure and the mass of air, sound vibrations could not reach our microphones. Because of the mass which air has, it causes a great deal of trouble to recordists working outside in windy conditions.

15

There is not much point in recording sounds we are unable to hear. So now we shall look at the nature of sound and the performance of our ears.

Frequency, wavelength and loudness

The human ear is very sensitive to small changes in pressure. Only when the rate of pressure fluctuations is increased to between 15 and 20 cyclic variations per second can the result be heard as a sound—a very low note.

The rate of change, in this instance of air pressure fluctuations, is expressed in hertz (Hz). 1Hz = 1 cycle per second, one cycle being a complete pressure wave cycle. 1kHz (kilohertz) = 1000 cycles per second. The velocity of sound-waves in air is 340m/s; therefore a frequency of 1kHz has a wavelength of 34cm and, likewise, a frequency of 10kHz has a wavelength of 34mm.

Such short wavelengths can present problems to the recordist. Where the reflected wave of a sound arrives at a microphone at exactly half a wavelength behind the direct-route wave (antiphase condition), cancellation occurs. Phasing, as these cancellations are sometimes called, produces variable sound quality.

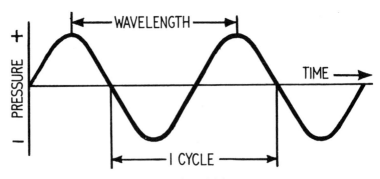

Fig 1 A sinusoidal wave

Phasing can be particularly troublesome when spaced microphones are used.

The range of frequencies perceived by the human ear is 20Hz to about 20kHz and is called the audio frequency (af) band. A frequency ratio of 2 : 1 is known as an octave. The ear responds to a range of nearly ten octaves. When a piano note is struck, a fundamental frequency and many harmonics (which are multiples of the fundamental frequency) are produced. It is the order and magnitude of the harmonics, together with the fundamental, which give the instrument its characteristic richness of tone—a richness which the recordist endeavours to capture faithfully on tape.

The sensitivity of the ear is not uniform throughout its total range. At 2kHz the range is 120dB (decibels) but at 30Hz it is only 80dB. Fig 3 shows the contours of equal loudness at different intensities of sound. The point where a sound is only just audible is called the threshold of hearing, and it can be seen that at 30Hz the sensitivity of the ear is 40dB lower (only 1/100) than it is at 2kHz. The sensation of loudness depends not only on intensity but on frequency; the phon, a measure of loudness, is used to distinguish loudness from intensity. The contours in Fig 3 have been arrived at by subjective tests in which the loudness of a tone is compared with that of a 1kHz tone adjusted to give the same apparent loudness; these contours therefore represent the loudness in phons at different frequencies and intensities.

When sounds at an original intensity of 90dB, eg an orchestra, are replayed at a much lower level, say 40dB, there will be a very noticeable lack of bass. This is entirely due to the poor low-intensity sensitivity of the ear at low frequencies. Therefore for listening levels greater or less than the original sound levels it is necessary to apply tone response corrections to the reproducing system in order to overcome scale distortion.

THE DECIBEL

Practically all measurement of sound-levels, whether acoustic or electric, is in decibels. When the output level of a circuit is divided by the input level, and similarly when a standard reference level is divided by a different level, a ratio results. Voltage and power ratios can be expressed in their logarithmic equivalents—decibels.

Shown here are some easy-to-remember values.

Decibels	Voltage ratio	Power Ratio
6	2:1	4:1
12	4:1	16:1

Decibels	Voltage Ratio	Power Ratio
20	10:1	100:1
26	20:1	400:1
40	100:1	10,000:1
60	1000:1	1,000,000:1

Power ratios, which as can be seen are the squares of the voltage ratios, have little significance in sound recording, whereas voltage ratios are commonly used. It may benefit the recordist to memorise a few common values. The convenience of using decibels will quickly be appreciated; they are additive and subtractive, whereas voltage ratios have to be multiplied and divided. The conversion graph on page 212 shows power and voltage ratios with their equivalents in decibels.

HEARING LOSSES

Returning again to the ear, it is an unfortunate fact that the sensitivity to higher frequencies decreases with advancement of age. At the age of 25 the upper limit may be 18kHz, at 40 years, 14kHz and at 60 years the upper limit might be down to 8kHz— and that is for 'good' ears. Long-term subjection to high-intensity sound levels is known to cause permanent loss of hearing. Reduced sensitivity to high frequencies is the most common impairment and the afflicted person does not hear speech with clarity, or music with full richness.

Another defect, which renders the ear almost completely insensitive to sounds at certain frequencies only, is caused by

Fig 2 Keyboard frequencies (fundamental)

exposure to high sound levels at any maintained frequency. This is known as notching. Motors and other sources generating loud whines or whistles can cause notching. Sources generating more than one loud, maintained frequency can cause multiple notching. When exposed to single-frequency noise for only an hour or so the ears become insensitive to that frequency for quite a long time after cessation of the noise.

The eye behaves in a similar way when subjected to over-bright light: when the iris cannot shut down the light sufficiently, the eye will desensitise. Most people at some time or another have experienced reduced visual sensitivity. For example, when returning indoors after sunbathing it seems almost dark for some time. Similarly, a temporary diminution of hearing caused by high noise levels is nature's warning that the dose received has exceeded the safety limit.

NOISE METERS

Sound-level measurement meters, ie decibel meters, are calibrated in two scales: uncorrected, for use by structural engineers and others to assess the probable adverse effects that high-intensity sound pressures may have on buildings; and weighted, that is to say corrected to a response which approximates to that of the human ear at different intensities, ie phons (see Fig 3). The second type is used to assess the damaging or likely annoyance value of sounds (noise).

The following weighted sound levels are typical:

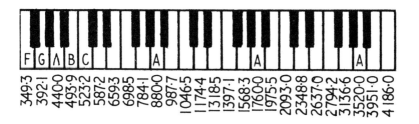

Threshold of pain	133dB
Jet aircraft at take-off	135dB
Pneumatic drill; underground train; noisy motorcycle	100dB
Loud radio; full orchestra	90dB
Average factory	70dB
Normal conversation	60dB
Average office	50dB
Quiet museum or public library	40dB
Whispering in quiet room	20dB
'Still of the night' in quiet countryside	10dB

How sound is modified by the carrier

The recordist will appreciate that sound levels may be very high, or very low. The level depends mainly on the subject and its proximity to the microphone, and to some extent on the sound reflected from adjacent objects. Sound-waves radiate spherically from a source, and hence the pressure decreases as the distance from the source increases. The rate follows the inverse square law, ie at twice the distance the sound pressure is one-quarter.

This is true for sound-waves in still air, but close to ground (or water) surfaces sound-waves follow the inverse square law less closely. High frequencies travelling closely over, say, long cotton-grass are attenuated at a higher rate, owing to the high absorption of the cotton-grass. Across still water, sound-waves are assisted by surface reflection. The attenuation is, therefore, less than the inverse square law suggests. Low-frequency sounds will 'hug' the ground surface (we have all heard of 'keeping an ear to the ground') and travel a much greater distance than would be the case in free, still air. The deep thud of ships' engines at sea can be heard for tremendous distances, as can aircraft engines being run-up at some distant airfield— neither of which, I might add, assists the recordist working outside. It must be remembered that in water sound travels at 1500m/s and could be assisting airborne sounds.

But by far the greatest modifier of the travelling sound-wave

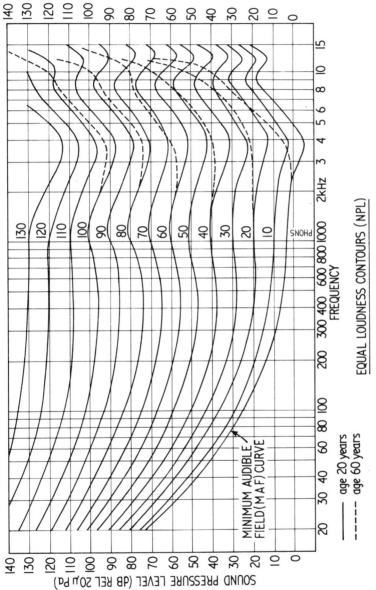

Fig 3 Equal-loudness contours (National Physical Laboratory; Crown copyright)

is the carrier itself—the air. The more turbulent the air, the greater the attenuation of sounds travelling through it. The sounds of distant trains are heard on days when the air is very stable, and are heard even more if the direction of movement of this air is from the source of sound towards the receiver. Motorway traffic is particularly noisy, as any roving recordist knows only too well, but when the wind direction is at 90° to the motorway and stable, on the windward side the traffic is almost inaudible at a distance of half a kilometre, whereas on the leeward side at that distance the traffic noise can be intolerably loud.

To the outdoor recordist the condition of the air, the sound carrier, is one of the most important factors to consider, for it can be either a major drawback or an asset. Air is very rarely dead still. When you take your recorder outside to record a street festival on a windy day, it becomes very clear that moving air can cause a lot of noise on a recording. Microphone work outside demands very good wind protection, and this is one of the topics dealt with in the chapter on 'Wildlife'.

2
Microphones

The microphone is the device which picks up sound-wave vibrations, converting them into electrical signals which can be amplified and used to modulate recording tape and to operate loudspeakers. The aim is to pick up, record and reproduce sound so that the result is a fair representation of the original.

There are many different types of microphones and in certain situations some may perform far better than others. If you are passing through a garden square with your recorder and wish to capture the sound of an arboreal disagreement in progress between two cats, and there is heavy noise from traffic circulating round the square, have you the right microphone to suit the conditions? A microphone which picks up sound from all directions equally would produce a very poor result. What is needed in this situation is a microphone which, at the same time as favouring the desired vertical sound, rejects the interfering horizontal noise. These exacting requirements are fulfilled by the use of a figure-of-eight response microphone.

There is not one sole characteristic response which makes one microphone superior to all others. Each type has a carefully controlled directional performance. The recordist can get the best from any given set of circumstances by exploiting these directional characteristics; the more familiar the recordist is with the different responses, the better his chances of good recordings.

To understand what microphones are, how they pick up sounds and how electrical efficiency determines their useful range, we shall look at the different types which are available. We shall also try to find out how manufacturers' technical blurb can help us select microphones.

Omnidirectional microphones
The omnidirectional microphone, *omni* meaning 'all', is the type

of instrument most usually supplied for use with domestic cassette and reel-to-reel recording machines. It comes in three basic forms.

CRYSTAL MICROPHONES

The diaphragm causes piezo-electric crystal salts to bend. Rochelle salt is the most commonly used material, owing to its high piezo-electric sensitivity. The bending of the crystals produces the electrical output. Another type of crystal microphone, the 'soundcell', has no diaphragm but is actuated directly. It has a flatter frequency response than the diaphragm type but is less sensitive. The output impedance (the generator electrical resistance) of crystal microphones is high and only short lengths of cable can be used between microphone and amplifier/recorder. Although the sound quality can be quite good, the crystal microphone has only one real asset: it is very cheap.

CAPACITOR MICROPHONES

These are of two types. The diaphragm in an electret microphone forms one half of a very small permanently charged capacitor; the sound-waves impinging upon the diaphragm cause the voltage across the capacitor to vary in sympathy with the sound. Because of the very high impedance of the capacitor the microphone has an impedance conversion transistor built in to give low to medium impedance output. The electret microphone is capable of good performance in regard to frequency response but gives a poor noise performance. Portable cassette machines often have electret microphones built into them, and here the noise picked up from the drive motor usually exceeds that of the inherent noise of the microphone. Built-in microphones are convenient but cannot be expected to give much in the way of quality sound.

The second type is the condenser microphone. This is perhaps better known as the 'studio' condenser. The active element is a capacitor, but it differs from the electret in having a much higher polarising voltage across it. The potential, usually of the order of 50V, is supplied either from an external source or from an oscillator-generator built into the microphone body. The inherent noise (hiss) level from a condenser microphone is usually very much lower than that from an electret one. But whereas an electret microphone is cheap, a condenser microphone is relatively expensive.

DYNAMIC OR MOVING-COIL MICROPHONES

In construction, a dynamic microphone is similar to a small loudspeaker with a handle. In an omnidirectional microphone the air-space at the rear of the diaphragm (or cone) is sealed to prevent sound-waves reaching it from the outside, although to relieve atmospheric pressure changes a very small air-leak to the rear is provided. The output impedance of a dynamic microphone can be supplied in any value by the use of integral transformers, but the more common impedances are 200Ω and 50Ω. Dynamic microphones are fairly robust and capable of good performance. The truly omnidirectional microphone is equally sensitive to sounds from all directions, but such microphones are difficult to come by. More usually one finds that they are omnidirectional at low frequencies only, becoming progressively directional as the frequency increases.

When I view a microphone it means nothing to me if the polar response is not known. When it is known, I can visualise imaginary lines of force round a microphone and these help me to position it. Also, in any set-up one immediately recognises the polar response most likely to solve a particular problem. It is just a matter of learning about the standard polar responses and getting to know a range of microphones and how their polar responses deviate from standard.

The *polar response* of a microphone represents its sensitivity to sounds from different directions. The sensitivity at different sound frequencies and at different angles of incidence can be measured and displayed on a polar response graph. A microphone with a polar response similar to that shown in Fig 4 (below) would have a useful pick-up angle of around ± 70° about the centre axis. Outside this angle there would be a substantial reduction in high frequencies.

An omnidirectional microphone is useful for some interview and effects recording. It has no directionally selective response; this can be a major problem when recording in noisy locations or where good separation is required between two sound sources.

25

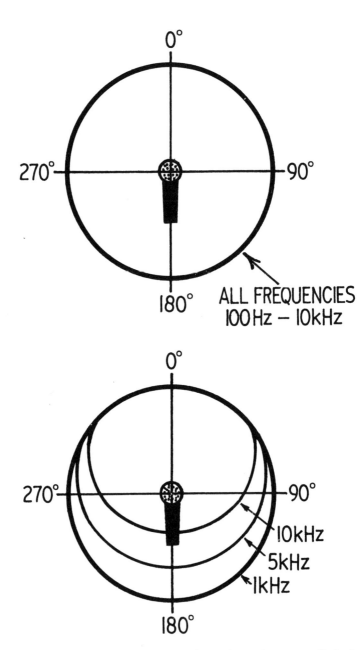

Fig 4 Omnidirectional responses: (above) a truly omni response; (below) the response more typical of omni microphones

Cardioid and unidirectional microphones

Cardioid microphones are most commonly either of dynamic or capacitor type.

A good cardioid microphone will have the same directional response at all frequencies within its range, whereas the directional response of the unidirectional microphone can be vastly different at different frequencies. *Unidirectional* means that the microphone is more sensitive to all sounds from one direction while the attenuation of sounds from other directions may be poor and frequency-dependent. A cardioid microphone is capable of 20dB attenuation at the rear—a voltage ratio of 10 : 1—and the recordist can take advantage of this to produce clearer recordings. He can select what to record and what to discriminate against by choosing carefully the position and direction of the microphone.

Hyper-cardioid microphones

Hyper-cardioid microphones can be of dynamic, capacitor or ribbon type. The output impedance is commonly 200Ω. A hyper-cardioid is more directional than a cardioid or, to put it another way, it is more selective: there is less pick-up of sound from the sides, and at about $130°$ around the centre axis there is a very high order of rejection. To achieve this, a hyper-cardioid microphone has a small pick-up lobe at the rear, the angle and sensitivity of which can vary a great deal, depending upon the design. In some the sensitivity at the back is half that at the front, while in others it may be only a quarter. The latter is generally referred to as super-cardioid. The phase of the rear lobe is in opposition to that at the front—a fact worth remembering when setting up a pair of microphones for stereo working. A hyper-cardioid microphone can be placed for maximum rejection of undesired sounds from one or possibly two directions, while favouring to the maximum extent the desired sound from another angle. This can be achieved in both the horizontal and vertical planes.

Figure-of-eight microphones

Microphones with a figure-of-eight polar response are mainly for professional studio use and are usually of ribbon or capacitor

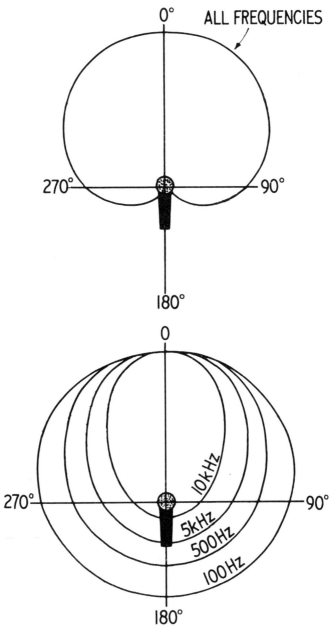

Fig 5 Cardioid (above) and (below) unidirectional responses. The unidirectional response is frequency-dependent

type. The studio condenser microphone is usually mains powered and switchable, either remotely or at the microphone head, to provide omnidirectional, cardioid, hyper-cardioid or figure-of-eight polar responses. Some recent stereo microphones of this type, designed to be powered from the microphone cables (phantom powering), can quite easily be powered from a batttery pack and are thus mobile. The modern electrostatic microphone, with a variable polar response, is an extremely versatile high-performance instrument and the high cost is merely a reflection of its merits.

Figure-of-eight ribbon microphones have been in use for a very long time, particularly in broadcasting, and their reputation as quality instruments remains untarnished. The STC4038 in particular is still regarded by many as the quality instrument by which others may be judged. A figure-of-eight microphone is equally sensitive at either face, although the phase relationship is in opposition. Therefore, sound-waves arriving on a plane with the ribbon (sideways on) will exert equal and opposing forces on the ribbon (or the capacitor element in the electrostatic microphone) and the electrical output will be zero. In fact, although the angle where the signals cancel completely is rather small, the 'plane-of-the-ribbon' rejection of a figure-of-eight microphone has to be heard to be believed. (This type of microphone could produce reasonably good results when used for the arboreal catfight.) I remember a test performed outside where there were no reflections, with someone speaking into the plane of the ribbon of an STC4038. The only sounds I could hear were coming acoustically direct from the person, and not from the microphone.

This brings to mind something which can catch out most recordists at one time or another. It is often perpetrated by high-spirited professional performers, and goes thus: the recordist asks for a touch of microphone balance/level, and the performer goes very quietly to the microphone and reads from his script in what visually appears to be quite a normal manner. In fact, only lip movements arc being made and there is no sound at all. Believing that what the eye sees must be true, the recordist progressively advances the gain of the microphone to a point where the landlord of that disreputable establishment in the next street might be heard putting kitchen scraps into the pig bin, at which point the performer says, very quietly, 'Do you want some level or something?' Have I been caught by that one a few times!

When ribbon microphones are used outside they must be given good protection against wind. The instrument contains an extremely delicate little strip of metal ribbon, suspended between two magnetic pole-pieces. One puff of wind is sufficient, it has been said, to tear the ribbon away from its mountings. I must admit I have not verified this—yet. Another point to remember when using a ribbon figure-of-eight microphone, is that the large, heavy magnet system is quite efficient at erasing recorded magnetic tapes. Imagine the anguish when, after

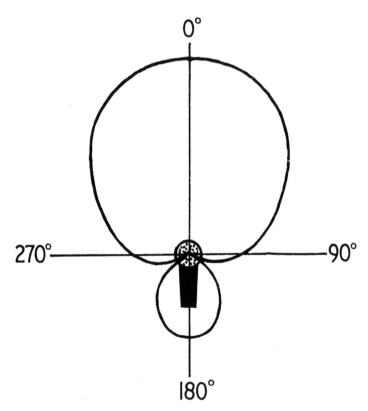

Fig 6 Hyper-cardioid response

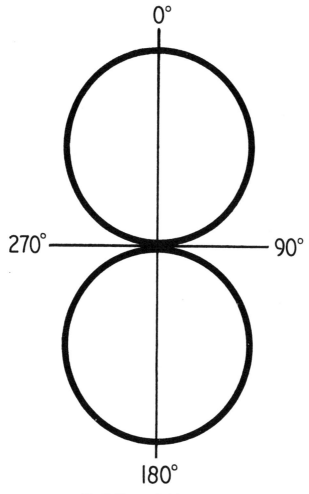

Fig 7 Figure-of-eight response

accepting the many kind offers of assistance to clear up following a choir-recording session, you arrive at the stacking point to find one or more large magnets lying immediately on top of the tapes just recorded!

Narrow-beam microphones

Narrow-beam microphones, more often referred to as gun microphones, have, as their name suggests, a narrow angle of acceptance. They are used by film-sound recordists who, because of the need to keep out of camera shot, cannot always get their microphone close to the subject being filmed. In tele-newsreel scenes, gun microphones are often seen in shot, though of course they may belong to another filming group. Gun microphones can be used to advantage (even with cassette recorders) by anyone requiring good, clean sound from a distance.

The principle of operation is the interference tube, which is a long tube with holes or slots along its length. These holes act as phase equalisers to control signals from outside the designed angle of pick-up.

As will be appreciated, microphones that utilise phase-reactive devices will exhibit frequency-dependent directional characteristics. The beam width can be as narrow as 30° at 8kHz, and as wide as 90° at 500Hz. Also, the output phase of off-axis signals can vary a great deal—often becoming negative—and this is a serious problem when working in stereo.

Line-source or column microphones

Another phase-reactive device is the line-source or column microphone. In the column microphone the on-axis position is sideways on—the reverse of the interference tube where the on-axis position is in line with the tube. The column consists of a number of spaced or stacked microphone units which are interconnected electrically.

However, the phase of off-axis sounds will vary at each unit, depending upon the angle of incidence and the frequency, as in the interference tube. It is obvious, therefore, that when the column is mounted vertically the polar response is narrow in the vertical plane and wide in the horizontal plane. This explains why the assembly is also called a conference microphone, for it is used at conferences and similar meetings, where it is ideal for picking up comments from the floor. The column is usually about 1.5–2m high, 150mm wide and 100mm deep; visually, it can easily be mistaken for a public-address loudspeaker column, which works on exactly the same phase wavefront principle, but in the reverse direction.

The column microphone is not a particularly high-quality sound-gathering device but, like the parabolic reflector, it can receive distant sounds very efficiently and in many circumstances it may be the only practicable system to use. The on-axis

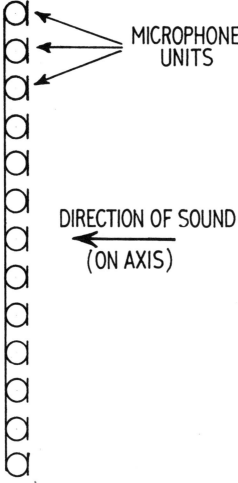

Fig 8 When sound arrives, at 90° to the column, the phase of the signal received by each individual sound unit is identical and the electrical output is therefore additive

sensitivity of a column is usually ten times that of a single cardioid microphone. The performance, incidentally, of a whole circle, square or cross of electrically coupled sensing units would be similar to, or better than, that of a parabolic reflector and would be more manageable than a reflector. The cost of the rather high number of units required in such a set-up is probably the reason why instruments of this kind are not in common use.

Pencil beam or parabolic reflector

A very narrow-angle polar response is achieved by using a reflector to concentrate incoming parallel sound-waves on to a microphone. A searchlight reflector is parabolic and it projects a narrow beam of light from a light source situated at its focus. The parabolic sound reflector will work with sound just as the searchlight works with light. It *can* be used to direct sound from the focal point and deliver it at force to some distance, but this is not the use to which the sound recordist normally puts the reflector.

A parabolic reflector has one important advantage over other sound pick-up devices: it is a noiseless acoustic amplifier. For a forward beam of 10° width a modest 600mm diameter reflector gives approximately 14dB (×5) gain for frequencies above 500Hz; a larger and less easily manageable unit of 1m diameter can provide up to 20dB (×10) gain to frequencies above 300Hz. The forward gain of a reflector is defined as the difference in output level of a microphone reflector-mounted to the same microphone unmounted. When the principal requirement is a good signal-to-noise ratio, nothing can better the combination of a sensitive dynamic microphone set in a large parabolic reflector. It is in the electrical amplification of signals that noises are generated. Therefore, the less the electrical amplification, the better the signal-to-noise ratio.

Reflectors are used only when it is impracticable or undesirable to approach more closely to the subject to be recorded. At sports and athletic meetings, football, cricket and polo matches, boat races and a host of other events, the sounds of the participants and their actions can be picked up on a panned reflector mounted quite high up and well clear of spectators and any other close-proximity sound sources. Undoubtedly, the reflector is a

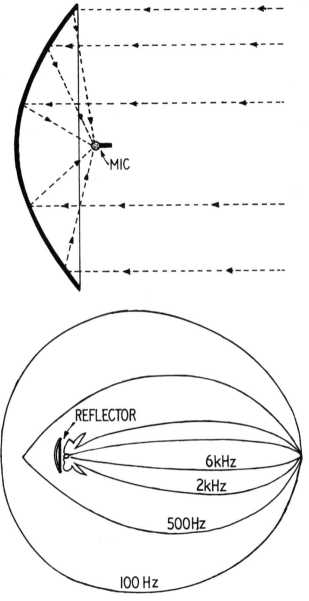

Fig 9 A parabolic reflector showing sounds being reflected to the microphone at the point of focus (above). Beneath it is shown the frequency/directional response typical of a reflector

valuable instrument. In wildlife recording it is probably the most popular device after the tape recorder, and its use will be discussed further in the chapter on 'Wildlife'.

Microphone data

Manufacturers make available data for every microphone, and many technical publications and magazines include microphones in their test reports. It is not sufficient to note only the brief specifications advertised in popular magazines: they inform one only that a particular microphone may have a cardioid polar response, a front-to-back ratio of so many decibels and a sensitivity of so many microvolts at a given impedance.

A lot more information is needed. We need to know the output voltage and impedance at frequencies from 100Hz to 15kHz and, particularly important for stereo, the polar response at different frequencies. For working near electrical apparatus we need to know the reaction of a microphone to magnetic and static interference. All these details are given in manufacturers' full data, usually available direct from the manufacturer and from approved agents.

FREQUENCY RESPONSE

The frequency response of a microphone may be seen quoted as 100Hz–8kHz. These figures, however, are almost meaningless until the parameters (deviations from a reference level) are also quoted. A frequency response of 100Hz–8kHz ±3dB, with overall response of 40Hz–14kHz +3dB −15dB, means that the deviation in output level, with respect to the output level at the standard test frequency of 1kHz (unless otherwise stated), is −15dB at 40Hz and at 14kHz. At 100Hz and 8kHz it is −3dB. Between 100Hz and 8kHz the variations do not exceed ±3dB from reference level. A microphone with this response would be suitable for recording some forms of music, many natural sounds, effects and all speech. A wide and flat frequency response may appear to be an important requirement in a microphone, but care must be taken to see that the bass response is not too far beyond that required for a particular subject.

Many microphones, capacitor types particularly, have a response down to almost zero frequency; sometimes this can be

36

very troublesome, for the very slightest air movements can cause rumbles. Even when a wind-muff is used, it is not easy to hand-hold or speak closely into such a microphone without producing the most awful pops and bangs on the output: the trouble given when zero-frequency microphones are used outside can be imagined.

Unfortunately, this is one of the characteristics which is rather underemphasised in microphone data, so take care not to select a microphone for its extended bass response alone when one with a sharp cut-off at about 40Hz might be more than adequate for your needs and relatively free from rumble troubles. Nearly all the sounds we are likely to record contain high frequencies, and the sounds of nature are predominantly so. A good high-frequency (hf) response is far more important than a good bass response. A fall-off in the bass can easily be corrected by additional amplification. It might bring up hum but this, usually a pure sine wave, can be neutralised if necessary by the addition of the same hum in reverse phase. Applying extra hf amplification to correct deficiencies in hf response increases hf electronic noise—hiss—and hiss is random-frequency noise, which cannot be neutralised.

All microphones have a source impedance, eg 200Ω. This is the generator resistance, and when a generator is terminated with a load resistance equal to the generator resistance the electromotive force (emf) produced will divide equally between the source resistance and the load resistance. In other words, the output level will fall to half the open-circuit level. Some manufacturers recommend a minimum load resistance for their microphones, but they cannot know what load is going to be applied; therefore, in all data the sensitivities specified are the no-load open-circuit voltages. These may be given in one of several ways: some manufacturers use millivolts, while others prefer decibels. The sound-pressure level used to derive the specifications can also differ. 10µ bar is popularly used but this may be expressed in the equivalents: 1 Pa (pascal), 1 newton/m² (N/m²) or 94dB. 10µ bar is about the sound level that an orchestra would produce when doing its best in a concert hall.

At concert-hall acoustic levels and above (80dB plus) the hiss noise generated by microphones is normally not obtrusive, because only moderate electrical amplification is necessary. However, when recording sounds of lower intensities, like stage plays, quiet speech and some of the many sounds of the outside world, where average levels are more likely to be around 50dB, electronic hiss can be very much in evidence and may sometimes be obtrusive.

Sensitivity is a most important factor to consider when choosing a microphone: nothing can make up for a microphone of poor sensitivity or poor signal-to-noise ratio, no matter what the polar response may be, because the more delicate sounds will be masked by high electrical noise.

The noise generated by a dynamic microphone, caused by thermal movements of electrons in the generator resistance and molecular movements of air against the diaphragm, is practically constant. Manufacturers' data give the measured noise from a 200Ω microphone as commonly around $0.2\mu V$. The maximum signal-to-noise ratio can be calculated for any microphone at any sound level. At 10μ bar a 200Ω microphone with a sensitivity of $0.25mV/\mu$ bar will have a maximum signal-to-noise ratio of $(2500/0.2)\mu V = 12500 : 1 = 81dB$.

Some electret microphones may not do quite as well. For example, take an electret type which is quoted as having a signal output of $5mV$ at 10μ bar and a noise output of $3.6\mu V$, and calculate the signal-to-noise ratio to be expected when recording quiet speech at about 50dB acoustic. At 94dB the signal-to-noise ratio $= (5000/3.6)\mu V = 1388$, approximately 62dB; at 50dB it would be $62-(94-50) = 18dB$, which would be totally unacceptable. Such a microphone can be used only where sound levels are high. Sometimes noise levels will be seen quoted in decibels, and 60dB might seem to be a reasonable figure at 10μ bar, but at sound levels around 50dB it would return a signal-to-noise ratio of 16dB, approximately $6 : 1$ as a voltage ratio, which would be intolerable.

Manufacturers' data often provide both weighted and unweighted noise figures or measurements and these can be useful in indicating what sort of noise predominates. When weighted and unweighted noise measurements are almost iden-

Maximum signal-to-noise ratios (S/N ratios) obtainable from a range of microphones

Microphone	10μ bar output 3kHz–10kHz (μV)	Noise (μV)	S/N ratio at 10μ bar (dB)	S/N ratio at 50dB (0.065μ bar)	S/N as voltage ratio
Beyer M69 200Ω	2400	0.2	81	37	70 : 1
Beyer M88 200Ω	3800	0.2	86	42	130 : 1
Beyer Sound-star 200Ω	2000	0.2	79.5	35.5	60 : 1
AKG D190 280Ω	2000	0.22	78	34	50 : 1
AKG D900 200Ω	2600	0.2	81.5	37.5	75 : 1
AKG C451 (CK8) 200Ω	1800	3.6	75	31	35 : 1
AKG C505 200Ω	5000	2.8	65	21	11 : 1
Grampian GC2 200Ω	1500	0.2	77	33	45 : 1
Grampian DP4 200Ω	4000	0.2	86	42	130 : 1
Grampian DP4 in parabolic reflector 200Ω	16000	0.2	98	54	500 : 1
Sennheiser MD421 200Ω	3500	0.2	84	40	100 : 1
Shure SM58 200Ω	2500	0.2	81	37	70 : 1

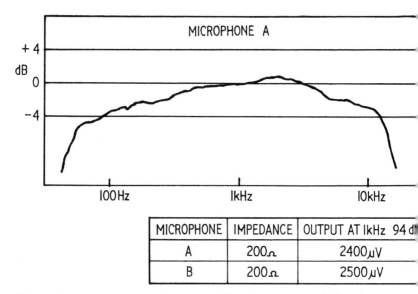

MICROPHONE	IMPEDANCE	OUTPUT AT 1kHz 94 dB
A	200Ω	2400μV
B	200Ω	2500μV

Fig 10 Output levels of two microphones. The rising response of microphone
    B produced a greater signal-to-noise ratio

tical, the noise is probably between 300Hz and 6kHz—the mainly flat portion of the aural curves. A small discrepancy would indicate mainly hf noise (hiss). A discrepancy of 12dB or more would indicate that the predominant noise was of low frequency—hum and rumbles.

One must also take the frequency response into account when dealing with noise, because nearly all the noise from a dynamic microphone is hiss (Johnson noise). Therefore, a microphone which has a rising treble response from 3kHz to 15kHz would return a better overall signal-to-noise ratio than the measured noise suggests.

Taking two microphones (A and B) we can compare hiss levels. Both are good-quality 200Ω dynamic instruments of practically the same sensitivities at 1kHz but, whereas the output of A remains flat through most of the response and falls off at 10kHz, the output of B rises to +6dB between 8kHz and 15kHz. Obviously the sound quality will be different but, when frequency corrections are applied to smooth out the responses, the hiss level of B will be found to be lower than that of A.

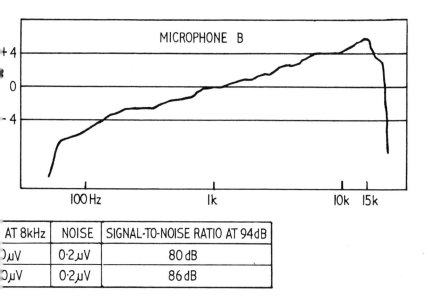

AT 8kHz	NOISE	SIGNAL-TO-NOISE RATIO AT 94dB
)μV	0·2μV	80 dB
)μV	0·2μV	86 dB

The Table of maximum signal-to-noise ratios (page 39) shows how the hf output level of a number of different microphones directly affects hiss level. Signal-to-noise ratios are given at 94dB and at 50dB acoustic and the hf output levels are assessed from published frequency response and sensitivity data.

Why do we need to consider the signal-to-noise ratio at low-intensity levels—surely it does not change with change in intensity levels? Indeed it does not alter, but then neither does the listening level follow, in volume, the intensity of the original sound. In fact, listening levels remain relatively constant irrespective of the sound being reproduced. Whether listening to a recording of a jet aircraft taking off at close range or to a recording of the sounds made by ants processing whatever lovely rubbish they process, we listen to them at levels vastly different from the original—at comfortable and convenient volumes. There is no loudspeaker anywhere which could reproduce the level of a jet taking off. A big steam-driven foghorn is the only device I know which is capable of shaking the dust out of a sailor's turn-ups at 20m. If we reproduce the sounds of ants at

41

original levels we do not hear them too clearly. To hear them comfortably we amplify the intensity level thousands of times. That is why we need to concern ourselves with low-intensity noise levels—if we intend to record anything other than pop groups or aircraft.

OUTPUT LEVELS AND IMPEDANCE

To allow for the various differences in test levels and units, all manufacturers specify a test reference level, and this may read: $0dB = 1V/dyne/cm^2$ (1 dyne/cm^2 is 1μ bar $= 74dB$). Where decibels are used in the data it is easy to calculate the level in millivolts. Take, for example, $-72dB$ ($0dB = 1V/dyne/cm^2$). If $0dB$ equals $1V$, then we know that $-60dB = 1/1000V$ ($1mV$); $-72dB$ is $12dB$ lower, or $0.25mV$.

The sensitivity of a given microphone is constant and independent of impedance, and a microphone gives different output levels at different impedances. The higher the impedance the higher the output, but not directly so. The impedance ratio is dependent upon the square of the voltage ratio. When output level differences are V_2/V_1 then impedance differences will be $Z_2 = (V_2/V_1)^2 Z_1$. The table of output levels and voltage/impedance ratios covers the range of most microphones and may be found useful when calculating for differences in impedance or voltage, remembering that allowance may be needed for any differences in sound test levels. It will be appreciated that at 10μ bar the output level from a microphone will be ten times that at 1μ bar.

To find out which of two microphones is the more sensitive—one of 200Ω with an output level of $-72dB$ or one of 600Ω with an output level of $-72dB$—we can refer to the table opposite to convert decibels to millivolts and impedance differences to voltage ratios:

Output levels: $-72dB$ (600Ω mic) $= 0.25mV$
$-76dB$ (200Ω mic) $= 0.16mV$.

Impedance difference as a voltage ratio is $600\Omega : 200\Omega = 1.75 : 1$. From this we can deduce what the 200Ω microphone would produce were it 600Ω:
$0.16 \times 1.75 = 0.28mV$ ($-71dB$)
—and what the 600Ω microphone would produce were it 200Ω:

Output levels and voltage/impedance ratios

Output (dB)	levels (mV)	Voltage/impedance ratio* Voltage ratio with respect to 200Ω	Impedance (Ω)
−50	3.3	2.2	968
−52	2.5	2.1	882
−54	2.0	2.0	800
−56	1.6	1.9	722
−58	1.25	1.8	648
−60	1.0	1.7	578
−62	0.8	1.6	512
−64	0.64	1.5	450
−66	0.5	1.4	392
−68	0.4	1.3	338
−70	0.33	1.2	288
−72	0.25	1.1	242
−74	0.2	1.0	200
−76	0.16	0.9	162
−78	0.125	0.8	128
−80	0.1	0.7	98
−82	0.08	0.6	72

$$*10:1 = 20k\Omega$$
$$15.8:1 = 50k\Omega.$$

$0.25/1.75 = 0.143mV$ ($-77dB$).

In this example, the 200Ω microphone is only marginally more sensitive than the 600Ω microphone.

200Ω has become the standard impedance for microphones, although other impedances will be found. 600Ω may still be found in some public-address equipment; high-impedance 50kΩ microphones are popularly used in domestic and amateur set-ups, where inputs are unbalanced and of high impedance. Whatever the impedance of a microphone it is important that it matches the amplifier input and the connecting cables. A 50kΩ high-impedance microphone connected via 100m of lighting flex to a low-impedance amplifier input would be unlikely to produce sparkling results! Transformers are often used for

43

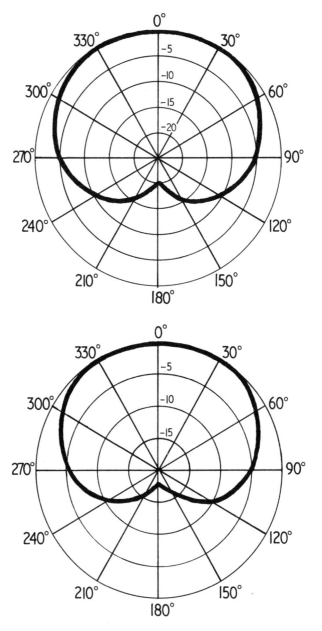

Fig 11 Identical cardioid responses but plotted on different dB circles

matching purposes and the impedance and voltage ratios can easily be calculated. A quick reference to the table of output levels and voltage impedance ratios will show that a transformer with impedances of 600Ω and 50Ω has a voltage, and turns, ratio of:

$$\frac{600\Omega}{50\Omega} = \frac{1.75}{0.5} = 3.5 : 1.$$

POLAR RESPONSES

Polar responses have already been illustrated, but do not be fooled when looking at manufacturers' plotted responses by the appearance of the shape alone.

The two cardioid polar responses in Fig 11 are identical but the one below appears to have a better front-to-back ratio than the one above. (The front-to-back ratio is the difference in sensitivities between the front and the back of a cardioid microphone.) The reason for the difference in appearance is that the diagram below has circles of sensitivity to only −15dB, whereas the diagram above has circles to −20dB. The −20dB diagram is the more common because a great many microphones have response points below −15dB, and some good cardioids have a front-to-back ratio of 20dB. 20dB represents a voltage ratio of 10:1. If we imagine that there is a constant-level sound source on axis (0°), the microphone will produce, let us say, 10 units; when the same sound source is swung round to the back of the microphone (180°) and remains at the same distance, the microphone will produce only 1 unit. With a 10:1 cardioid the visual appearance of the polar diagram below in Fig 11 is more representative of the volume/distance differences that are found in practice. Some manufacturers now use a unit system for plotting microphone data.

Fig 12 shows the difference in appearance between a plotting on the decibel circles and the same plotting on the unit circles. The plotting on the unit circles informs us much more clearly that at 90° the sensitivity is one-half of that at 0°.

When the polar response of a microphone is non-uniform, ie the response varies with frequency, the sound quality of off-axis sources will be different from the quality on-axis and may vary quite appreciably, depending upon the angle of incidence. The

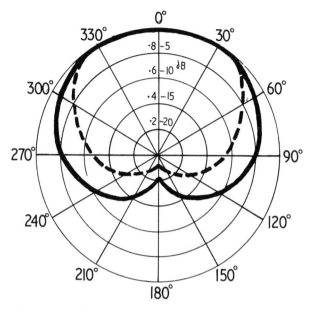

Fig 12 Identical cardioid responses plotted on (solid line) dB circles and on (broken line) unit circles

most likely deviations from uniformity are that the high-frequency response is narrower than the middle-frequency response, and the low-frequency response is rather wide. When a microphone with these variations is used for speech and the person speaking wanders off-axis, or the angle of the microphone is varied, the result is a change in voice quality—similar in effect to that experienced when a person speaking without a microphone turns bodily from side to side. When the speaker faces the listener the full quality of 'presence' is heard, but as he turns off-axis his voice loses this quality; the more off-axis he turns, the greater is the loss in quality and intelligibility. Thus, the less its response deviates, the better the microphone.

'Shotgun' or 'rifle' microphones exhibit quite large deviations in directional responses: as they stand, the forward angles of pick-up at frequencies below 1kHz are, at best, inferior to good hyper-cardioids. Improvements are continually being made, and perhaps one day a shotgun microphone with a nearly uniform polar frequency response will be produced. I am not saying that current types are poor, because they are not: they have good forward selectivity. It is a microphone which has many merits, and one which I use frequently.

Another way to illustrate the polar responses, and one often used to supplement the sensitivity circles, is to show the responses at different angles on the frequency response graph, as in Fig 13. The amount of unevenness of the displayed response depends on the degree of damping used on the pen recorder. A high damping will result in a smooth line, whereas if little or no damping is used the response line can be very uneven and represent more faithfully the response of the microphone under test.

Considering the high cost of most microphones, a dealer may be reluctant to lend an individual a number of microphones for a few days for experimental purposes. Even if a selection was made available, very few people have access to facilities for properly assessing the performance and suitability of a microphone for their needs. It could take weeks before the advantages of one microphone over another becomes apparent and, furthermore, how many potential buyers know just what they are looking for?

The informed prospective buyer will first gather the manufacturers' full data for the microphones under consideration, knowing that the difficult test work has been done and that the published performance data represent a true statement of the results of such tests. Having studied the data and decided on

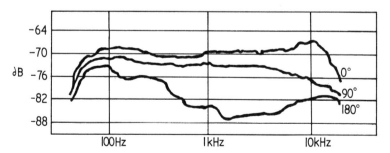

Fig 13 Directional response of a microphone displayed on a frequency-response graph

possibly no more than four microphones from the hundreds available, the buyer is in a better position to ask a dealer for the microphones on a sale or return basis. Do not allow a dealer to sell you a microphone he especially recommends if there is insufficient data available to support the recommendation. The dealer could be right, but how are you to know that the recommendation does not originate from simpler motives—like profit margins!

On numerous occasions I have used some very popular, high-quality condenser cardioid microphones with interchangeable heads and adjustable angle-joints and they have produced a quantity of excellent stereo recordings of birds in close-up, much radio drama and special effects, as well as location talking-with-action recordings. Unfortunately, these little microphones have proved to be rather unreliable. Used outside, especially as the sun goes down and the dew-point changes, I have found them to whistle, pop, crackle, hiss and buzz and become unusable.

So, if you are considering buying a capacitor microphone make certain that it continues to work satisfactorily when the working temperature and dew-point change. If necessary, ask the dealer to prove that the microphone works properly in adverse conditions by, for instance, placing it in the ice-box of a refrigerator for ten minutes. When the microphone is removed from the ice-box it will be below the dew-point, and moisture will quickly condense all over it and possibly inside it. Check that it works and continues to work for an hour or more after removal. If it does not, either reject it or buy a heater coil to keep the microphone warm!

There is another range of capacitor microphones which give nothing but perfect performance in the most adverse weather conditions imaginable. I have the greatest confidence in these no matter what, or where, the task lies. They would probably continue to work after total immersion in water, but considering the very high cost there has been an understandable reluctance to verify this supposition. The reliability comes, I am sure, from the superior modulation technique in which the active capacitor element forms part of a radio-frequency (rf) circuit at very low impedance, whereas the other type has the active capacitor in a high-impedance audio-frequency circuit. The rf microphone to which I refer, and have used a great deal, is the MKH815 made by Sennheiser.

I found, however, that the instrument easily overloaded when used in situations of high sound intensity. A colleague of mine was not aware of this and used a gun microphone to record the sound of the first Concorde ever to take off from Bristol. He was recording some distance from the runway but was still shaking some time after the beautifully shaped machine had disappeared beyond the horizon. The recorded result, although exciting, was nevertheless very distorted. I made the same error when recording birds on an island. A pair of Sennheiser MKH805 gun microphones were placed close to a colony of sandwich terns and I remember thinking at the time that the sound level was rather high—a very harsh and piercing shriek—so why, I wonder, did I use the gun microphones when others were available! We learn by our mistakes: the calls from the birds nearest to the microphones were distorted and had subsequently to be edited out.

The distortion arises in the microphone because above a certain sound pressure the internal radio-frequency modulator and demodulator circuits are driven beyond their limits of linearity. In the data for the Sennheiser MKH816T—the latest of their 12V operated gun microphones—the maximum sound pressure level is 118dB. With sound at that level, is a gun microphone really necessary?

Incidentally, a degree or two of distortion can heighten the drama of some events and sound effects. If a really lively crack of thunder is required for a stage or radio play, one way to achieve it is to superimpose over the very front of a fairly normal crack-bang-rumble a very sharp and grossly distorted version of the same crack-bang. It makes recorded thunder sound just that little bit louder, nearer and more dramatic. The tell-tale tape hiss, which often warns a listener that something is about to happen, should be cut off. Shock sound effects should be preceded by silent, non-magnetic leader tape. The effect of thunder treated in this manner is so dramatic that it should not be played at high volume to those of a nervous disposition without prior warning. Anyone who has experienced in the quiet of an early morning in the country the shock of an unseen propane-operated bird scarer going off at close quarters will know what I mean.

In this chapter we have looked at microphones and data and have seen that sensitivity is a most important factor when recording sounds of lower than concert-hall levels. With a good understanding of the different directional performances we are better able to select the right microphone for a given situation.

A recordist who is familiar with microphones can produce far better recordings, even on a cassette machine, than a person who makes recording machines his priority and ignores the fact that quality depends on how he uses microphones.

Later we shall study the various methods employed, and the types of microphone used, to obtain first-class recordings. Next, though, we shall investigate the causes of extraneous noise on our recordings.

3
Noise

If there is a recordist who has never been troubled by extraneous electrical noises, I have yet to meet him. Any sound recordist who boasts that problems have never come his way has obviously not done much recording or has had unbelievable luck. From time to time every recordist encounters noise problems: the novice with his cassette recorder connected to the radio may meet hum; the professional with his multi-channel mixer may pick up spurious signals from fluorescent lighting, television sets and radio transmitting gear. For every noise problem there is usually a simple answer. If we study the different types of interference and why they occur we shall better understand how to design or arrange our recording gear to be interference-free—or at least to work out procedures to rectify any troublesome noises.

The main noises to examine are: hum, static crackles, sizzles and buzzes; radio-frequency interference; and hiss. These can all present interesting problems from time to time.

Electrical hum
Low-frequency noise, which can vary from a smooth, deep rumble to a spiky buzz, can be caused by a number of factors. The commonest are: unbalanced, unscreened and damaged cables; dirty, corroded and broken connectors, causing high resistance and imbalance; moisture or condensation in the connectors, causing spurious leakage currents; microphones of incorrect impedance for the cables or amplifier inputs; microphones of low sensitivity, necessitating high amplification, which magnifies any hum caused by other factors; poorly screened microphones, causing static interference clicks and spurious hum leakage currents when handled or placed in position; microphones which are susceptible to interference from static and magnetic fields; amplifying equipment linked incor-

51

rectly to the electricity mains, causing multi-path earth currents; mains-powered equipment with poor dc smoothing, or poor electrical or magnetic isolation from the mains.

In most domestic and semi-professional equipment the microphone inputs are of high resistance and suitable only for high-impedance sources; one of the pair of signal conductors is connected directly to the amplifier chassis and possibly also grounded, leaving only one live (and unbalanced) conductor to carry the signal. In respect of interference pick-up and distance working, this is poor in principle.

High-impedance microphones should not be used with long cables, because the cable will exert a damping effect on the high-frequency signals. Five metres of cable can load a microphone with 500pF (picofarad, ie 10^{-12}F), and at 10kHz that is a load of 30kΩ. When this load, together with that of the amplifier, is placed across a 50kΩ source, its output level will fall by about 9dB. A length of 100m would drop the level by 30dB at 10kHz.

If it does become necessary to use high-impedance microphones with long cable runs, impedance-converting transformers will be needed. One transformer should be inserted between the microphone and the long cable, to convert the high impedance of 50kΩ to a low impedance of around 200Ω to send to line. To avoid earth currents and static interference, the output to line should be electrically isolated from any other connections. At the recording or receiving end of the cable, the output can be taken directly into the microphone amplifier, where this is a low-impedance, balanced input, or fed to another impedance-converting transformer between the cable and the amplifier, if the amplifier has a high-impedance, unbalanced input.

Anyone considering using long cables with high-impedance input cassette (or open-reel) machines and matching high-impedance (or low-impedance) microphones will gain enormously from the use of impedance-converting transformers. Many microphone manufacturers list transformers in their data and information sheets; they are available as loose items with fly-leads (for mounting into equipment) and as complete accessory units in a variety of small cans and boxes and with a variety of input and output plugs. These units are simply plugged between microphone and cable, cable and recorder or anywhere else where impedance conversion is necessary.

An alternative to the use of transformers would be to amplify the signals at the microphone end before sending them through the long cable. An amplifier of around 40dB gain would be required for this and the input should be a good impedance match with the microphone. The output impedance of the amplifier should be as low as possible—200Ω or less. Amplification of low-level signals before sending them to line is the conventional method used to distribute programme and communications signals by wire. The sound recordist is unlikely to operate at distances greater than 1000m from the source and little or no trouble should be experienced when a few simple principles are adhered to.

MICROPHONE CABLE

Hum, radio break-through and various forms of static induction interference are minimised by using the correct type of cable. The cable designed for the job is microphone cable—a tough, plastic-covered, twin-conductor cable with an antistatic screen—and this should be used for all microphone circuits irrespective of the distance involved. There are several types available, including cables designed specifically for use in areas of high static interference, for example in television studios, where lighting cables and apparatus radiate strong interference fields.

The two more commonly available types are: standard cables of about 6mm diameter (preferred by most professionals engaged in day-to-day recording duties, because of their greater resistance to damage); and lightweight cables of about 4.5mm diameter (used more by those engaged in activities where weight is a concern and where long cables are the normal requirement and have to be run out at high speed).

The lightweight cable requires a greater degree of care in handling and laying than does heavyweight, for it is very easily damaged by, for example, trampling feet—especially on rough and uneven surfaces. I have made many a desperate dash to reach, in advance of farm and forestry vehicles, lightweight cables lying across a rough track and thus save them from certain ruination. Even the toughest of cable would be put to the

53

test by a spike-wheeled tractor pulling a disc roller! If it is necessary to lay cables across a track which is expected to carry some traffic, it is advisable to scratch a groove across the track, placing the cables in this and covering them with sand or fine earth. Make sure that the filling material contains no sharp stones, otherwise the exercise may prove futile. If no filling is applied the cables will easily ride out of the groove; loose stones on the track will be thrown around by passing vehicles and may fall into the groove to damage the cable when the next wheel runs over it. If the outer cover is damaged and the screening conductor is exposed, it should be repaired with a moisture-repelling insulation material.

EARTH LEAKAGE CURRENTS

Earth leakage currents between any points of contact with the ground and the recording point, small as they may be, can be transferred to the signal conductors and hence be amplified with the signal. On long cable runs, earth leakages can result in pick-up of radio signals, as strong rf earth currents may be induced in the cable. Also, long cables make good radio aerials. On various recording ventures outside I have received short-wave radio stations and also marker beacons, the latter giving interrupted carrier only and no audio modulation. Radio beacon signals applied to some high-gain microphone amplifiers can cause some very strange effects on the local desired signal: intermittent suppression and screeches of instability are just two examples. Earth leakage currents can produce many undesirable clicks, sizzles and hums for, no matter where, the earth as well as the air is loaded with signals ranging from sub-audio frequencies to radio frequencies.

On one site where I spent three weeks, and where it seemed to rain almost continuously, the microphones, fifteen in all, had to be left outside for the duration of the exercise. There was a great deal of trouble with earth leakage currents. All the microphones had to be electrically insulated from contact with the ground and any vegetation—and as this was at a badger sett it was not easily achieved. Trouble also came from the large number of connections en route. These filled with condensation and rain-water, although well bound with insulation tape and covered with plastic bags. To locate the problem spots, it was necessary to communicate to 'base' via a walkie-talkie while checking the joints for condensation. As fast as I dried and re-insulated one joint another would be found to be giving trouble. We had so much trouble that in the third week of recording (also the week of nightly transmissions on BBC television) a special multi-circuit cable was installed from the sett to the control point, some 300m distant. Thereafter, only occasional trouble was experienced, caused by a microphone being in contact with garlic leaves or some other vegetation, or by moisture in a microphone connector.

These problems show how important it is to avoid all earth leakages along a cable run. Had it been possible to install a multi-microphone pre-amplifier at the sett, most of the troubles would probably have been insignificant, although water does tend to reduce signal levels somewhat!

Having seen what pitfalls to avoid, let us now consider what is required for a trouble-free set-up. The screen of the cable should be connected directly to the chassis of the microphone amplifier/ recorder and to the body or screen connection of the microphone, thus screening the whole system from static interference. The microphone amplifier should have a balanced transformer input which is a good conversion match for the microphone. The microphone should be of high sensitivity, of low to medium impedance (200–600Ω) and have the best polar response for its intended purpose. If a metal-bodied microphone is connected directly to the screen of a cable, ensure that the clip or mounting device which holds the microphone is insulated from the ground or the microphone stand. Most microphone clips are now plastic and insulated, and metal ones usually have rubber liners.

When the conditions as set out in Fig 14 are adhered to, cable runs of up to many hundreds of metres can be used with confidence. It is important to note that the output of the microphone must at all times be balanced. Many dual-impedance (200Ω/ 50kΩ) microphones are not balanced: the output cable is only single conductor plus screen and the impedance change is affected by repositioning the output connector so that different output pins are connected to the cable. Their output is from auto-transformers in the body of the microphone; because of the

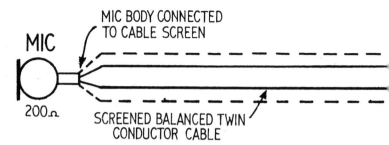

MIC BODY CONNECTED
TO CABLE SCREEN

MIC

200Ω   SCREENED BALANCED TWIN
CONDUCTOR CABLE

Fig 14 Connecting a balanced-output microphone to a balanced, twin-conductor screened cable and balanced-input amplifer

need to ensure that the microphone body is connected to the cable screen and to the amplifier chassis, to achieve complete static screening, the cable screen, which is one side of the audio connection, has to be connected to the body of the microphone.

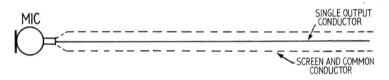

MIC   SINGLE OUTPUT
CONDUCTOR

SCREEN AND COMMON
CONDUCTOR

Fig 15 An unbalanced-output microphone

STATIC INTERFERENCE

An unbalanced output and cable cannot be used at normal microphone output levels much beyond a few metres without some interference pick-up. If an unbalanced microphone is connected to a balanced twin-conductor cable it must be connected to the conductor pair only, leaving the screen of the main cable to provide merely an antistatic screen to the cable pair. The problem with using an unbalanced microphone comes when handling it, for then the electrostatic balance is further thrown out and static interference results. To minimise static, the conductor carrying the signal from the body connection side of the microphone should be connected also to the chassis of the microphone amplifier.

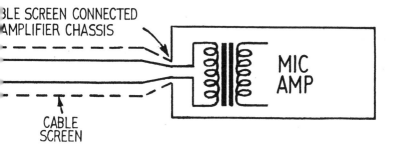

LE SCREEN CONNECTED
AMPLIFIER CHASSIS

MIC
AMP

CABLE
SCREEN

This converts the normally balanced transformer input of the amplifier to an unbalanced one; this may give more trouble than the occasional static resulting from handling, because the static screen of a cable carries a fair amount of interference, some of

Fig 16 Unbalanced-output microphone used with a balanced cable and balanced-input amplifier

which gets electrostatically induced in the signal pair. When the pair is electrostatically balanced the induced interference is equal in strength and phase in each conductor and is self-cancelling in the input transformer. This is the main reason for having balanced inputs.

RADIO-FREQUENCY INTERFERENCE

It will be found that different makes of microphone cable vary greatly in their susceptibility to rf interference pick-up. Even with good cables and amplifiers, trouble may occasionally arise because of dirt, rust or corrosion of contacts on the cable run. Dirt and corrosion can cause rectification of radio frequencies (just as in a crystal set); the resulting audio signals are fed directly into the input of the microphone amplifier and receive substantial amplification, no matter how much rf suppression is incorporated in the amplifier.

When I was a small boy I used the wires around fields to communicate with my friend some distance away. We both had high-impedance headphones and were able to communicate with one another over considerable lengths of field wire. One day I connected my headphones to a rusting field wire and received the (then) Home Service of the BBC, loud and clear!

When trouble is experienced with radio signals, it is wise first to try cleaning all the connectors on the cable run by plugging and de-plugging them several times until they are 100 per cent electrically clean. If that does not cure the trouble then the problem lies elsewhere and further investigation is required. Of course, you could sit back and enjoy the programme.

Microphone amplifiers can cause rf signals to be rectified and amplified. Even simple transistor amplifiers have responses up to many hundreds of kilohertz, which is most undesirable in audio amplifiers used for normal recording work. Not enough attention is given to the suppression of radio frequencies in audio equipment generally. Some designers seem to revel in the fact that an amplifier designed for audio responds to many megahertz. Perhaps when the designers of these amplifiers actually try to use them in locations heavy with radio-taxi activity and bleeper call systems, they will be less impressed by the results of their super-wide-band amplifier designs.

There is an obvious need to design a bandwidth limit into audio equipment. Microphone amplifiers in particular, should receive special attention. It is in the early and sensitive stages of an amplifier that rectification of rf signals occurs and causes trouble; therefore, rf signals must be suppressed—before they receive amplification. Usually, this can be achieved by inserting an rf choke and capacitor between the input transformer and the first amplifier stage.

The following diagrams will give some idea of just how effective an rf choke and capacitor can be, and show the results obtained when a resistance replaces the rf choke.

To ensure that rf signals are not radiated to any of the electronics in an amplifier, the transformer and the rf suppression

FREQUENCY	IMPEDANCE			RESPONSE
	L = 5mH	TRANS	C≈470pF	∂B
20kHz	700 Ω	5KΩ	18KΩ	−3
200kHz	7KΩ	5KΩ	1·8KΩ	−18
2MHz	70KΩ	5KΩ	180 Ω	−52

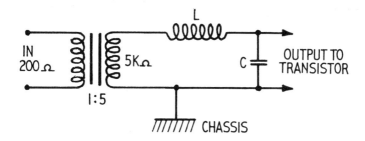

FREQUENCY	IMPEDANCE			RESPONSE
	R=2KΩ	TRANS	C=470pF	dB
20kHz	2KΩ	5KΩ	18KΩ	−3
200kHz	2KΩ	5KΩ	1·8KΩ	−14
2MHz	2KΩ	5KΩ	180 Ω	−32

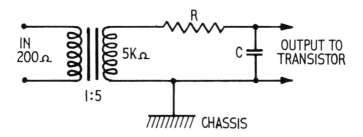

Fig 17 Suppression of radio-frequencies. Transformer losses at rf have been ignored, so the results will be much better than those shown

components should be enclosed in an electrostatically screened compartment close to the amplifier first stage. It is not enough to rely on total suppression at the first stage, because not all amplifiers are going to be perfectly screened; therefore, rf suppression should be incorporated in every stage.

The suppression performance of an amplifier can easily be checked by connecting the bare end of a small length of insulated wire to various components. The wire (2m is usually more than enough) will serve as an aerial to pick up rf signals, and when applied to a poorly suppressed amplifier the results will be alarming: the amplifier may even go into oscillation. This is quite a severe test, of course, but on a well suppressed amplifier the most that should happen at any point of contact is a bit of crackle as the contact is made, and perhaps some af static from fluorescent lamps and television receivers. A 20cm length of wire is enough to cause some poorly suppressed amplifiers to blurt out the local sports results; considering that the majority of stereo record players have half a metre of some of the most ineffective screened lead imaginable, is it any wonder that the local radio ham is picked up by a few of them? Unscreened loudspeaker leads are a source of potential rf pick-up; they act as aerials injecting rf signals into an amplifier's negative-feedback network.

Hiss

Recordings with high background hiss are unpleasant to listen to, irrespective of their content or general sound quality. Recordings need not be hissy. Noise-reduction systems are quite common on cassettes (otherwise inclined to be rather hissy), and reel-to-reel recording, if at reasonable speed, should be fairly low in tape hiss. So, why is there a hiss problem? Is it because the microphone is poorly matched to the recording machine or because the microphone amplifier is noisy, or what?

The problem of hiss is complex. It is due to thermal movement of electrons in the circuit resistances of amplifiers; the more one amplifies, the higher the hiss level. In other words, if one begins with an amplifier of good design—one in which the self-generated noise is very low—the main source of noise generation is the very first resistance in the circuit—the microphone.

60

Every microphone produces a certain amount of noise along with the desired signal. We have also seen that sensitivity and impedance are factors which determine the signal-to-noise ratio of a microphone. It would appear that once the quietest high-gain amplifier and the most sensitive microphone are combined in operation, nothing can better the resulting signal-to-noise ratio. We do not need to improve on it anyway, unless we intend to record sound levels below about 60dB, because we have seen that at 10µbar the signal-to-noise ratio can be as high as 80dB with a good dynamic microphone. When allowances are made in the listening level for sounds of an original intensity of 60dB, the reproduced noise level will be around −50dB, a little better than the noise from the average noise-reduced cassette recorder. When we come to record sounds of a very low level we find that with listening-level compensation the signal-to-noise ratio gets worse.

The amplifier circuit shown in Fig 37 in Chapter 9 has a high gain and low self-generated noise. The balanced-input transformer steps up the microphone signal to the first-stage transistor by ten times (200Ω:20kΩ). This arrangement gives the best performance with regard to signal and noise. Looking for a moment at the input, and replacing the 200Ω microphone with a 200Ω low-noise resistor, we can measure the full-gain noise level at the output; this will be comparable to the hiss level of the 200Ω microphone buried in a pit of sand to exclude all airborne sounds. When the 200Ω resistor is replaced by one of 100Ω the hiss level falls to almost half the value. When the input is terminated with a short circuit, the hiss level is very low compared with that from a 100Ω resistor and can be taken as the self-generated noise level of the amplifier.

The output voltage from the amplifier would be about 1V at a sound intensity of 0.1µbar (54dB), remembering that a sensitive 200Ω dynamic microphone supplies only 25µV at 0.1µbar. It will be appreciated that if the impedance of the source can be made very low the best noise performance is to be expected. Unfortunately, the use of microphones of a lower impedance results only in a lower signal level, so there is no advantage there. However, if more than one microphone of the same type is used to drive the amplifier, it can be seen that while the output level

61

increases, the impedance decreases.

Here, then, we have the solution to better signal-to-noise ratios. When two 200Ω microphones are connected in parallel and phased positively, the signal-to-noise ratio improves by more than 6dB with the 92dB amplifier. Taking this a stage further, four such microphones could be used in a series-parallel arrangement to provide a possible improvement of 12dB or more over a single microphone. The heads of the microphones would have to be very close to one another and in line (vertically) to avoid high-frequency phasing, which might otherwise reduce the improvement.

Weight, size and cost must all be considered in the drive towards improved signal-to-noise ratios when recording low-level sounds. Multi-microphone systems are further discussed later.

The MKH815 rf capacitor gun microphone manufactured by Sennheiser has an output impedance of only 20Ω (and an unusually high output level); when connected to the amplifier mentioned above, the noise from the amplifier alone was negligible. The result on a signal level of 50dB was checked against that obtained when using paralleled Beyer M88 microphones with the low-noise amplifier. The MKH815 was found to be only marginally more hissy than the paralleled M88 arrangement.

Acoustic noise

In this section we have taken a fairly serious look at electronic noises but little has been said about acoustic noise. In the previous chapter we saw that the use of a microphone with the appropriate polar response enabled us to counteract some environmental noises. However, we may sometimes place a microphone to record something and find that there is an environmental noise in our headphones—a noise we cannot hear when the headphones are removed.

This may prove to be more a peculiarity of the surroundings than of the microphone, for some quite baffling problems often arise concerning noise. For example, a sea cove with cliffs may, due to its scale and contours, act as an efficient sound reflector at low frequencies while absorbing or dispersing higher frequencies. A problem which may arise when recording in a cove is that noises from things beyond the horizon (aircraft or ships) can be picked up by a microphone placed at the focal point of the cove by a recordist who is unaware of its acoustic idiosyncrasies. The monitoring point may be located some distance from the microphone(s) and in a position where the noise from the unseen source is inaudible. The recordist may therefore be quite puzzled as to the origin of the noise. Stone quarries and similar locations should be viewed with a degree of caution too, for while many such locations will provide quiet shelter from environmental disturbances, some may actually amplify sounds from distant sources sufficiently to cause recording difficulties.

4
Recording Machines

Cassette and open-reel machines

The advantage of cassettes is their simplicity of loading and the ease with which any number can be pre-set for use, for example, in a lecture. With reel-to-reel tapes this would normally be done with markers in the tape to indicate the different bands to be played; the process of changing the sequence of the bands once they have been assembled becomes a slow and sometimes embarrassing task, whereas changing a cassette takes little more than two seconds. The cassette machine is unlikely ever to equal the reel-to-reel machine when it comes to recording/replay quality and noise levels: as techniques in electronics and mechanics advance to improve the performance of cassette machines, so they are applied to reel-to-reel machines as well.

The recordist is concerned first with the actual recording of sounds in the most realistic and faithful way possible and only after that with the leisurely but critical playback. Therefore, consider carefully the total facilities which may be required. A cassette machine may be ideal for playback and for dub-editing listening tapes from 'originals' and 'masters'. Battery-operated, portable stereo cassette machines are obviously favoured by the mobile recordist. To be independent of the mains is a great advantage, for then recordings can be made wherever it is possible to carry a machine. Many recordists may prefer reel-to-reel recording machines because of the many extra facilities available. I have known interviewers who have chased a good 'hot' story and recorded the interviews and any supporting sound effects (actuality, as they call it), then composed and recorded the vital narrative material on a portable reel-to-reel machine, editing the out-of-sequence material into a presentable item while returning to base.

The professional or keen amateur wants to be able to cut-edit the recorded material—to snip out a word or phrase here and there. Where a need for editing exists, systems are usually evolved to perform the operation. On cassettes, this could be achieved by release clips on the cassette which allow the two spools to be removed for editing, or by an automatic withdrawal of the tape from the cassette. However, editing would be rather difficult and perhaps inaccurate, owing to the very low running speed and narrow width of the tape.

The output from a recording tape is directly related to its speed and track width. A stereo compact cassette has four separate recording tracks on a 3.28mm tape. Each channel receives its signal from a 0.6mm head width. Compare this with the 2.5mm head width on a half-track reel-to-reel tape. The differences in magnetic track widths together with the differences in tape speeds, of between 4.75 and 19cm/s, can give large differences in frequency response and noise.

Taking published data we can compare the performances of two machines listed by a well-known manufacturer. For the comparison we shall select a cassette machine described in the blurb as the ultimate in high performance, and a high-performance reel-to-reel machine. Both are in the domestic/semi-professional bracket.

It is probably true to say that reel-to-reel 9.5cm/s half-track stereo (without noise reduction) is comparable to noise-reduced cassette stereo. Certainly, if the figures in the table of machine performance (see below) are anything to go by, we can see that

Comparison of machine performance

Characteristic	Cassette (with Dolby noise reduction)	Reel-to-Reel (without noise reduction)
Speed	4.75cm/s	19cm/s
Track	2 + 2	2
Frequency response	30–1300Hz ±3dB	30–20,000Hz ±3dB
Noise	−53dB	−56dB
Distortion	2%	1.2%
Wow and flutter	0.1% rms	0.06% rms

the performance of the reel-to-reel machine is superior in every way; if the noise-reduction system employed in the cassette machine were employed in the reel-to-reel machine the noise level would be even lower, around −66dB.

Nevertheless, compact cassettes will undoubtedly be of interest to most recordists, if only as a means of easy storage and quick selection for listening. There are many machines with good noise-reduction systems which are well worth consideration. There is, for the field recordist in particular, a need for a lightweight high-quality portable stereo cassette machine, but the currently available machines with noise-reduction systems are as large and as heavy as professional reel-to-reel portables. A quality cassette machine is still physically larger than, for instance, the Swiss Stellavox SP8 high-performance professional portable reel-to-reel machine. The cassette mechanism is small but its small size has not been exploited. In fact, there now seems to be a trend towards fitting these small mechanisms into cabinets of ever increasing dimensions, thus losing on two counts—quality and size.

The domestic compact cassette system may not greatly improve; perhaps the introduction of the small-based cassette (not the miniature) will prove to have been only an expedient during the current obsession with miniaturisation. The time must surely come, though, when cassettes will be able to cope with the operational requirements of every broadcasting and professional organisation. These may not be the compact domestic cassettes, but they will certainly be based on them, though running at higher speed and having a wider track width. Several manufacturers are now producing machines suitable for professional users and these run at 9.5cm/s, using a tape width of 6.35mm. The cassette is about twice the size of the domestic version and is as easy to use.

To get some idea of the performance of such cassette machines take the EL7 Elcaset produced by Sony as an example. It has three motors and three heads, permitting off-tape monitoring. It has a frequency response of 20Hz–25kHz ±2.5dB, a signal-to-noise ratio of 62dB and wow and flutter of 0.04 per cent—remarkable considering the 9.5cm/s tape speed.

The 33rpm record has been in use so long because it meets the demands of professionals, amateurs and domestic users alike. The 78rpm record, once so popular, has gone and reel-to-reel tape is now used only by professionals and keen amateurs. In practice the compact cassette (owing to its performance being below that which is acceptable in professional circles) is only used for domestic listening and recording. Most equipment designed for the professional can be downgraded for the domestic market, whereas it is less usual for domestic equipment to be upgraded for the professional. However, the Elcaset is, in principle, an upgraded version of a very popular and widely accepted domestic standard. When it in turn becomes slightly downgraded—if it has to be downgraded at all—to revolutionise the insatiable domestic market, the original form of compact cassette will probably fall out of favour.

The super-quality standard just has to come—and soon. For the moment, however, the recordist has little choice but to pay the price for a quality machine, or take chances which the professional could not afford.

Compact cassette machines and the lower-cost reel-to-reel machines have no off-tape monitoring nor even good level-metering, and their recording quality cannot be expected to approach that of professional machines. Nevertheless, very satisfying recordings can be made even on cheap machines, and the whole purpose of this book is to encourage recordists to enjoy making recordings and to make better ones, using whatever equipment they may have.

Recommendations for a particular tape recorder cannot be given, because when it comes to quality and serviceability one gets only what one pays for. It is for the buyer to decide on the degree of sophistication he requires and what he can afford. However, a few general points about tape recorders and recording tapes are worth looking at.

Performance
FREQUENCY RESPONSE

A response of 20Hz–20kHz is very good but by no means unusual, and it would be quite pointless to reject a machine only because of a slightly inferior bass response. Choirs, soloists, almost any sound ever to be uttered by a human, and practically all wildlife sounds, contain no frequencies lower than 75Hz; and only a limited number of musical instruments, among them the organ, grand piano and double bass, are capable of producing frequencies lower than this.

The characteristics of the listening environment also affect low-frequency response. The listening-room and loudspeakers may both be incapable of reproducing frequencies as low as 20Hz, 30Hz or even 60Hz—and only very good and usually rather large loudspeakers are capable of faithfully reproducing frequencies below 60Hz. Except for a few musical instruments, recording does not demand an extended bass response, but practically all the sounds one is likely to record—especially wildlife sounds—demand a very good high frequency and transient response. If either of these is poor anywhere in the chain, from microphone to final reproducer, the sound will be unrealistic—it will be lifeless.

For most work an adequate record/replay response would be 100Hz–15kHz ±2dB with a tolerance between tracks on a stereo machine better than 1.5dB. The signal-to-noise ratio will be dependent upon the type of recording tape used, and the noise contributed by the recording and replay amplifiers should be well below the lowest tape noise. 60dB is reasonable for recording and replay amplifiers and 55dB is a good overall signal-to-noise ratio for a tape speed of 19cm/s. With the incorporation of good noise-reduction systems figures in excess of 70dB are normal.

WOW AND FLUTTER

The term wow is used to describe the slower variations of speed and pitch up to about 5 per second (a 33⅓rpm disc with an eccentric centre hole will wow at 0.55 times per second). Faster variations are termed flutter and these effectively modulate the material being reproduced. The flutter frequency of cassette machines and 19cm/s open-reel machines is commonly around ten times per second.

Cassette machines, which have hitherto been rather poor in flutter, are, with improved standards of engineering, becoming acceptable in this respect. The most noticeable flutter frequencies are those of 25 per second and above. The amount of wow and flutter that can be tolerated will depend on the material being reproduced and on the listener. Not all listeners are endowed with equally critical aural perceptions and a percentage of flutter which passes unnoticed by one person may be unacceptable to another. Other factors which determine the acceptable wow and flutter are the material being reproduced and the flutter frequency. The piano is an instrument which can present a severe test for any recording machine, particularly for one whose flutter frequency is fairly high, say about 40 per second. The organ presents a severe test with regard to low-rate wow, where pitch variations on sustained notes are easily noticed, but a degree of low-rate flutter may be quite unnoticeable, as it is similar to the accepted 'tremolo'. The critical ear will detect flutter on piano music when it exceeds about 0.1 per cent rms.

TRACK STANDARDS

Careful consideration should be given to track standards. The professional standard is half-track stereo. For reasons of tape economy, quarter-track is very popular with the amateur—cassettes are quarter-track. Any recordist intent on producing master tapes to professional standard will consider tape economy of minor importance and will use the half-track two-channel standard. Half-track enables the talented enthusiast, the professional and even the novice to the art, to cut-edit for blemishes, for time and for insertion of coloured item-identification leaders, and to offer the prepared master-tapes to commercial enterprises. On quarter-track two-channel double-run machines (double-run being the only reason for using quarter-track) editing is not practicable, because a cut made in material running in one direction directly affects the material running in the opposite direction, and quarter-track is commercially unacceptable.

THREE-HEAD MACHINES

These facilitate off-tape monitoring while recording. This is a very useful and most desirable feature, affording an instant quality comparison of the recorded and incoming signals. With many three-head recording machines it is possible to switch from a 'play only' mode to a 'record/play' mode while the tape is running, thereby making dub-editing possible. When blemishes and stop/starts have to be cut-edited out, much time can be wasted if differences in level, balance or quality occur as the edit join runs through, thus necessitating repeated re-records and cuts.

Editing out faults when originals are copied to masters is child's play with good dub-editing facilities. It can be heard immediately when a dub-over works, and if it does not, little time is wasted in running back and re-dubbing until an acceptable edit is achieved. Some machines may produce a sharp audible click as the dub switch-in occurs, but this is not too serious if it is convenient to cut-edit out the clicks subsequently. Master tapes should not contain 'sticky old cut-edits'. If a machine is required to perform perfect dub-overs for mastering, it must not produce extraneous noises. Intending purchasers of a three-head machine would be well advised to ensure that it has facilities for dub-editing and performs it well (see also 'Dub-editing' in Chapter 10). Unfortunately this capability is not commonly referred to in manufacturers' data, and inexperienced sales staff may be unaware of it. Therefore, if in doubt, try it out or seek the advice of the manufacturer or his agent on this point. Cassette machines can be used for dub-edits, as can two-head open-reel machines, but it is necessary to spool the tape back to the dub point in order to ascertain the success of the dub-over; this can be very time-consuming if, after a long follow-on dub, the result is found to be unacceptable.

AZIMUTH

The alignment of the recording and replay heads on a tape recorder is very critical. Each head has an extremely fine gap in the magnetic material forming the pole pieces over which the tape runs, and these gaps must be absolutely at right-angles to the path of the recording tape. Azimuth is the term used to describe this critical alignment.

Fig 18 shows a correctly aligned azimuth and the resultant in-phase signals from each track. Above is shown an incorrectly set azimuth with sloping magnetic bars and it can be seen that the phase relationship of the output signals is in opposition.

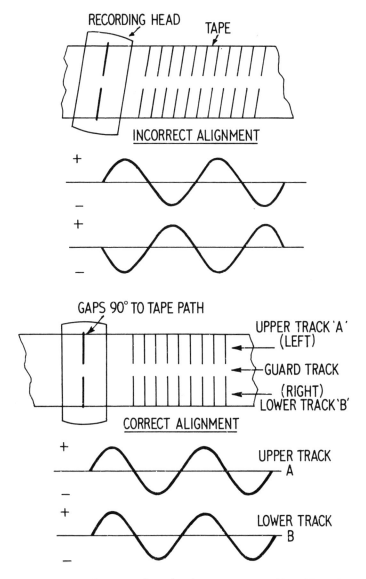

Fig 18 Azimuth alignment of tape heads on a stereo machine. Above, incorrectly adjusted azimuth and the opposing output waveforms resulting. Below, correctly aligned azimuth and resultant in-phase waveform

They will cancel each other when played on a correctly aligned head switched to A + B mono output, or will produce out-of-phase signals in stereo. The degree of cancellation or phase error will depend upon the signal frequency, the alignment error and the tape speed—the higher the tape speed the lower the error—which is one good reason for using reasonably high tape speeds. Azimuth errors produce a degradation of the treble response—especially on tapes played in mono. Azimuth alignment should always be done using a full-track (mono) tape of correct azimuth, for it is far more important to have the two tracks on a stereo machine in-phase to the highest usable frequency than to align individual tracks for maximum treble response.

When a two-head machine is used for playing back the material recorded on it, there should be no phase errors at all, however far out of alignment the head may be, the one head being common to both record and replay. There is one proviso: that the hf bias to each track is also correct. An ac bias of between 50 and 200kHz is applied to the recording heads, with the audio signal, to imprint the audio on to the tape in the most efficient way. Without ac bias the recorded signal would be distorted and weak.

BIAS

There is a critical bias for each type of recording tape and for different recording speeds. As the ac bias is increased on a recording, the phase of the signal magnetic flux is caused to lag very slightly behind the magnetic head gap and this lag primarily affects the phase of higher audio frequencies. Consequently, when there is an imbalance in the hf bias there may also be phase differences in the recorded audio signals. Each type of tape has a recommended bias level and the adjustment is usually done while recording a specified frequency—normally a steady tone of 10kHz. The bias is taken from zero (at this point the recorded 10kHz signal should be distorted and weak) to a point where the signal reaches a maximum level; the bias is then increased still further until the recommended over-drop in signal level is reached—commonly 3–4dB. And, of course, all the tracks on a machine are treated in the same manner.

The bias adjustment controls are usually mounted inside a

machine and are accessible on removal of a panel or on removal of the machine from its case. Adjustment will be somewhat more difficult on a two-head machine, as it may require recording and replaying several times to set the bias correctly. The difficult part will be in finding the point of maximum signal level. Once that is found the over-drop can easily be set. When the bias is increased beyond the point of maximum signal the reason for the drop in recorded level is that, like the hf signal applied to the erase head for erasing any signals on a tape prior to a new recording, the bias partially erases signals at the recording head as they are being recorded. Sound-on-sound recording, remember, partially erases the higher audio frequencies of the original material.

There is another method of applying bias: it can be applied to a separate 'bias-only' head mounted at a slight angle directly opposite the recording head (which then carries the audio signal only), the tape running between the two. This system of biasing, generally called cross-field, is claimed to suffer less partial signal erasure than does the more conventional direct method (where the signal and the bias are fed together) and consequently the signal-to-noise ratio is better.

EFFICIENT WORKING

For a recorder to work at maximum efficiency it is important to ensure that the recording and replay heads are free from deposits of tape residues and are magnetically clean. Tape residues, which build up amazingly quickly, can be removed with a soft cloth and a spirit cleaner such as methylated spirit. Not only should the tape heads be cleaned frequently, but the guides round which the tape travels should also be kept scrupulously clean.

MAGNETISED TAPE HEADS

A replay head with a degree of permanent magnetism will cause the weaker and more delicate parts of the reproduced signal to sound rough or distorted (sounding similar to low-level sounds reproduced from an out-of-centre loudspeaker cone), owing to the dc bias given by the magnetised head. The dc bias would cause low-level signals to be suppressed in either the positive or

73

negative half-cycle at a point determined by the strength and polarity of the bias.

A recording head with some permanent magnetism would cause the signals applied to it to be offset magnetically. The hf bias would thus be unbalanced and distorted; this would cause audible noise (rather like the roar of the sea at some distance) and the audio signals would reach tape-saturation level (distortion point) earlier. But probably the worst effect of magnetised-head recordings is noticed when the tape is physically cut, as when cut-editing and leader-inserting, for it cannot be cut without producing a sound like a dull thump. Cuts break the continuity of the magnetic bias and this causes a stronger flux across the cut. The result is even worse when dc biased tape is cut and joined to unbiased tape, and worst of all when two tapes having opposing dc biases are joined together. When faced with problems of this nature, the only reasonable solution is to copy the faulty tapes, using clean machines, and use the copies for cutting. The dc bias will not be carried through to the copies.

Recording tape

Different makes and types of tape exhibit vastly different hiss levels, and differences in sensitivity and maximum modulation levels. Tape hiss stands out prominently in recordings of a low general modulation level, as in some recordings of the sounds of nature where sounds may occur only once in five or more seconds.

Mid-frequency tape noise is quite effectively masked by ventilation noise in studio recordings and by atmosphere in recordings of natural sounds. When the recorded level is continuously high, most tape noises are masked. To understand how masking works, consider for a moment the small sounds made by pins falling into a wooden tray, these sounds being interrupted intermittently by the sound made by a hand-bell. We hear all the pins hitting the tray up to the very moment the bell rings; then our attention is suddenly diverted from the original sound and we hear the bell only. As the sound of the bell diminishes we begin to hear the pins once again. If the sound of the pins is replaced by a similar sound, or if it changes in level the same instant that the bell rings, the chances are that the change in background will not be detected. Masking is used extensively in tape-editing to lessen the impact of a change in background level.

Tape hiss is a function of the tape, not of the recording machine. If the same piece of tape is used for recording on a variety of machines, the hiss level will be similar. Different machines may vary in the maximum undistorted level which can be recorded on a given tape, because of different bias settings or (on some machines) special antidistortion circuits.

The problems involved in selecting recording tape are almost as complicated as those in selecting microphones but there is one advantage: tape is a lot cheaper. Tape manufacturers will supply full technical data on available tapes, but the simple guide leaflet is sufficient to show that there are many types: thick, high-modulation; thin, low-noise; dynamic; super; extra-dynamic; high-output, low-noise. To confuse things further there are matt-back, polished-back, normal, extra, double-play and even triple-play tapes. To go into all the differences in detail would require another book, but the essential differences are outlined here.

Matt-back tape is preferred by professionals because it has a very much smoother rewind characteristic than polished-back tape, and this is very important. Permanent damage can result from unevenly wound tape: the thinner the tape the higher the risk of damage and the more severe when it happens—and it can happen simply by picking up an unevenly wound reel of tape and allowing the fingers to push over and crease projecting layers.

PRINT-THROUGH

Thin tape cannot usually carry high modulation level without excessive print-through. This is caused by a section of highly modulated tape coming into contact with both the previous and following layers of tape, as naturally happens when it is wound on to a reel. It imparts some of its information to these layers by direct magnetic influence. The audible result is that pre- and post-echoes are heard; there may be printing through several layers, causing multiple echoes on either side of the principal sound.

GRAIN SIZE

The grain size of the magnetic coating material affects the noise content of the recording. A fine-grain coating may give a very low hiss level but not necessarily a good signal-to-noise ratio. For example, if a low-noise tape gives half the hiss level of a high-output tape, but the high-output tape can support four times the signal level, the signal-to-noise ratio of the high-output tape is twice that of the low-noise tape.

DROP-OUT

This is the term used to describe a complete or partial reduction of the recorded signal. Drop-outs may be only a few milliseconds in duration but can be very noticeable, particularly when listening to stereo recordings, where sudden positional shifts may occur. These variations are due to irregularities in the magnetic coating material. Over-biasing will help to reduce hiss and distortion and also lessen the effect of tape drop-outs. Under-biasing will increase high-frequency efficiency at the expense of more distortion and drop-outs of greater severity.

CHOOSING TAPE

Without being swamped in a sea of data how do we find out which tape is right for our machine? With a new machine the manufacturer usually provides a list of suitable tapes, although should the manufacturer also produce his own recording tapes it is unlikely that other brand names will be mentioned. Many magazines on tape-recording and related topics do test reports on recording tapes and machines, and the information given often covers important points not mentioned in manufacturers' data. I remember working with one brand of tape which was difficult to cut without causing the magnetic coating to part company with the backing. A fault found with another brand was a lack of straightness: the tape rode up and down as it passed the tape heads. The term 'weaver' was coined to describe this fault and about 60 per cent of the tapes supplied at the time had to be rejected.

Obviously these 'weavers' and 'shedders' are the result of manufacturing errors. The visual appearance of tape on a reel is no guide to its quality: this is found out in use. Independent test reports are helpful, but one cannot assume that because a brand has a high reputation and good test reports there will not occasionally be some inferior tape inadvertently released for sale.

A colleague found that the yellow marker tape he was using to divide items on a reel could be recorded on—not quality recording but, nevertheless, it did hold modulation of a sort. However, when the polished surface of this marker tape was facing the recording heads it gave no problems at all; this is probably the best way to use marker tape when cutting it into matt-back recording tape.

In the hope that we do not come across too much inferior tape, let us progress to a few simple tests of comparison. It is assumed that the experimenter will be able to obtain a small length of each type of tape considered for test. One way to compare performances of a number of open-reel tapes is to cut-edit together small lengths of each type (10–15 seconds of each should be enough) and wind them on to a reel. Run the machine on 'record', with no input signal, through to the end of the last sample (it is as well, also, to turn the recording control to zero). Then run the tape back and play it through with the volume/ replay level up high, and make notes of the hiss level from each type of tape. It may be necessary to rearrange the order of some sections to be certain of the hiss level, so that each one is progressively quieter (or noisier). Each section of tape can be marked on the back with a laundry marker so that its identity is not lost.

The second test is to find out how much signal each sample can hold. Again, run the tape on 'record', but this time record a constant level of speech from a high-quality source, eg a VHF radio. The object of this exercise is to find the distortion level of each sample. It may be necessary to run through the tape several times, advancing the recording level on each run. After noting the order of tape saturation levels, record another run-through using some very bright-toned music (eg piano and violin), keeping the level below that which causes distortion on any section. Play back and note the order of treble response.

To test for print-through, record a single hard-struck piano note at a level just below the distortion point of each section. Record this note about half-way through each section in turn and replay the whole tape, noting the level of pre- and post-echoes on either side of the note proper. Leave for an hour or so and recheck.

To test tape for drop-outs, it will be necessary to obtain enough tape for about two minutes running. Constant-level modulation such as white noise should be used for this test. A VHF radio, when not tuned to a signal and not inter-station muted, will provide a suitable noise. The recorded noise should be played back at reasonable level in order that sudden changes in quality or level can be heard. These will be most noticeable when replay is via a stereo system, with the listener in a 90° position between the loudspeakers. In this position tape drop-outs will be very obvious to any listener with normal hearing. When using a radio as a noise source be certain that any negative markings are not caused by interference pick-up on the radio.

If a metering system is available the findings should have an added degree of accuracy, but do not be a slave to meters. We use ears to listen with, not meters, and if our judgement does not agree with meter readings are we to question the meter or our judgement? I have more confidence in my trained ears than in meters, although I use meters a great deal for reference levels.

Similar tests can be carried out on cassettes, but the time interval necessary for cassette changes might make it more difficult to assess the order of merit. Cassettes should also be tested for smoothness of running: they should not bind or drag or spool erratically. After the full run of tape assessment a picture should have emerged of the most satisfactory tape to use.

Noise-reduction systems

Tape noise can be reduced by the use of reduction systems such as the Dolby or DBX systems. In general, noise-reduction systems increase the level of low-level signals during recording and reduce them when replaying. There are other systems—Philips' Dynamic Noise Limiter is one—in which the high-frequency noise (hiss) from the tape can be totally suppressed on replay. The system has a control for adjusting the threshold so that it may be set just above tape hiss level and only signals above the threshold are allowed through. It is obviously more effective to boost low-level signals before recording and subsequently reduce them on replay, as the dynamic range is then improved. In the threshold system the dynamic range is limited by the tape noise, and if the tape noise is about −55dB the dynamic range cannot be much more than 52dB. This is quite adequate for the more constant sounds like pop music, where there is hardly a moment when the modulation level falls to anywhere near tape-hiss level, but when listening to wildlife sounds, where the atmosphere level can be low but at the same time so vital, the results may not be too good.

Good noise-reduction systems can add a new dimension to a tape recorder, but care should still be taken to select the right tape for the machine in order to reap the full benefit.

Tape speed

Unfortunately, the better noise-reduction systems add considerable extra weight and bulk to a machine and this may rule out their use when mobility is essential. It might be better to use a higher tape speed than to carry extra weight. A recording made at 4.75cm/s requires eight times more amplification on replay than one made at 38cm/s and results in about eight times the noise level as well as eight times the trouble with drop-outs. Reasonably high tape speed has the advantage of giving a better frequency response. The tape dilemma can probably be summed up as being performance versus cost.

On the Nagra 4 and Sony TC510/2 machines which I use, a 19cm/s tape speed has been found most satisfactory. High-modulation, matt-back tape of good overall performance is used, but to get an even better signal-to-noise ratio from the tape I boost the high frequencies by 4dB at 10kHz on recording. The amount of boost to be applied will depend largely on the capabilities of the machine; it is worth experimenting to find how much can be applied on your particular machine before the onset of distortion on the tape. The boost applied during recording will cause the recorded material to sound too 'bright', so during playback it is necessary to remove the emphasis with frequency-correction circuitry—and this is where the benefit is found because, as well as removing the emphasis, the correction also reduces tape hiss. In fact, as far as tape hiss is concerned, I find that the result is comparable to that at 38cm/s normal. A

19cm/s tape speed is used mostly in mobile field recording where small reels of tape are the norm, whereas 38cm/s is the standard speed in studio recording.

Off-tape monitoring

I have explained at some length that the right microphone, the right tape for the machine and reasonably high tape speeds are fundamental requirements for high-quality recording, but I have not yet mentioned much of the operations side of the art. To ensure that the last drop of signal-to-noise ratio is extracted from the tape, in a technical sense, it is essential to modulate the tape to the maximum. On a recording machine with off-tape monitoring it is a simple matter to ensure that maximum level is applied to the tape. It involves no meters or anything at all technical—just the keen ears of the recordist: wearing a pair of high-quality headphones or listening to high-quality loud-speakers the recordist should advance the recording level until distortion is just perceptible, and then hold back just below that point.

When the machine has no off-tape monitoring it is not poss-ible to tell where the distortion point is. Only *after* recording can the tape be played through, and then one can only tell if the tape is over-modulated (by the distortion present) and not whether the level could have been higher without distortion. Recordists using machines lacking off-tape monitoring will nearly always play safe and under-modulate the tape to avoid any danger of distortion, and in doing so will forego the possibility of a better signal-to-noise ratio. An over-modulated tape is a recording ruined—with one exception: when recording wildlife sounds it is quite in order to over-modulate grossly an unwanted species which happens to be too close, in order to obtain a good signal level of a more distant, desired species. Cut-editing (or dub-editing) will later remove the distorted and unwanted sounds, leaving the desired sounds clear.

Automatic recording-level systems

If carefully designed and correctly set for the type of tape in use, these can overcome the problem of over-modulation and can also help to prevent under-modulation. In a good auto-control

80

system the general level will be set by the first signal of more than half a second duration and this level will be held for at least 10 seconds, after which, if no other signal approaches that level, the gain increases slowly to maximum. In addition to a general level control there will also be a limiter with attack and decay times of around 20μs and 200ms respectively, which operates over the main auto-level and deals with short-duration peaks of high-level modulation which would otherwise escape and cause distortion. Such systems are capable of very good results—far better than can be obtained by a half-awake operator; even an alert operator cannot react as quickly as an electronic limiter.

I have modified the limiters on my Sony TC510. They were not limiters but auto-level controls having a recovery time of over 10 seconds. They were also unganged, and that was unacceptable in a stereo machine. The limiters are now ganged and have three different switchable recovery times: 100ms, 600ms and 8s. The attack times of the first two are extremely fast and the auto-level (8s decay) has an attack time of approximately 400ms. These simple modifications were very easy to effect. The machine microphone inputs are unbalanced and low-gain so I use a special microphone amplifier on the line input. With the gain turned up high and with limiter 1 or 2 in operation I can shout into the microphone, delivering ten times normal signal level into the machine, and it does not distort. Now, that's what I call a limiter!

A good operator will anticipate a burst of high-level signal and reduce the recording gain before it occurs, but the problem is to know just how far the gain control needs to be moved to accommodate the sudden higher level. The experienced recordist will not be able to tell you how he gauges this, any more than a cyclist can say how a bicycle is ridden, except to say that with practice it becomes almost automatic and somehow the level on the tape turns out about right. If limiters are used (and only prejudiced purists would not use limiters) they should be set to operate only on what the recordist aims to limit electronically. They should not override the responsibility and skill which the recordist ought to be exercising, any more than a climber should rely on a safety line—the climber controls the line and the line should not be allowed to influence the climber's intentions.

Automatic recording

Automatic recording can be very useful in many situations. Intermittent sounds—those sounds which occur once in a while and after a long wait (and when they do the operator is reading a book and thereby misses the sound entirely)—can be recorded using a sound-level switch which automatically switches on the recording-machine and controls the level. The machine thus started usually continues to record for a period after the cessation of the sound which set it in motion. Sounds recorded on automatic machines often require some work done on them in a copy/transfer stage to iron out level irregularities.

With sound-switch recordings there is an additional problem: the recording has no lead-in atmosphere and starts on or very slightly after the start of the desired sound; often there is an obvious wow-start which will need considerable treatment to produce listenable results. There is a way to overcome this problem and obtain sound-switch recordings which start a long way before the sound begins, but it involves the use of a continuous recording-loop on another machine. One machine, the one to which the signal is fed, is left to record continuously. When the required sound triggers the sound-switch, the second machine is started and this receives its signal from a specially mounted replay head positioned in advance of the erase and recording heads on the loop machine, so that the sounds existing on the loop, and which contain the lead-in sounds up to and including the required sounds, are recorded on the sound-switch machine.

Recording-level meters

Recording-level meters come in two types: Volume Unit (VU) and Peak Programme Meter (PPM). A PPM has a linear scale (in decibels), as the unit is driven by a logarithmic amplifier. A VU meter has a logarithmic scale and is merely an alternating-current voltmeter. The display can be shown electronically by light-emitting diodes, or mechanically by pointer meters. In the mechanical type there are vast differences in the design of the meter movements. The PPM has a special 'fast-to-register-without-overshoot' movement and is superior in every way; hence the cost of a PPM movement is far higher than that of a VU movement.

PPMs are fitted as standard in professional equipment and VU meters (because they are cheaper) are fitted to domestic and semi-professional equipment. VU meters may be satisfactory for recording organ music but they are certainly no use when trying to record bird calls and similar sounds of short duration because, as the meter is trying to indicate the average level, the true peak level may by that time be running into severe distortion on the tape. The scale of a VU meter is logarithmic: all the low-level signals are crushed up at one end, with the last 6dB occupying half the meter scale; the needle spends most of its time thrashing about from one end of the scale to the other, which even a sober operator will find hard on the eye and pretty meaningless.

Many cassette machines and open-reel machines now have mechanical metering systems which, although nowhere near as good as a PPM, indicate the peak level of the signal as long as the peak is of sufficient duration to activate the electronics in the meter-driving amplifier. These peak-holding meters do not act in such an abandoned manner but actually assist recording. There is no reason why the electronic meter should not register perfectly, whether PPM or VU peak-holding, but in the mechanical type a very fast movement meter is very costly and this is the reason why it is not standard in semi-professional equipment.

Headphones

Fieldwork has been mentioned several times but in this book it has no connection with ploughing! It means that one is not tied down to a particular set of circumstances, as is the case in a studio environment. It need not mean only headphone monitoring, for loudspeakers can be used almost anywhere: in cloakrooms, passageways, half-way up a lighthouse, a thousand metres below ground level in a tin-mine, in a car, and even afloat on a rowing-boat. For field recordists with no static monitoring facilities, headphones will be the normal means of quality assessment. Headphones are much more candid as monitoring devices than loudspeakers will ever be, but the user will need to be experienced in headphone listening to make a critical judgement in stereo and advanced systems of recording.

What does the field recordist require from a pair of (stereo) headphones? At first, high quality might seem to suggest high cost but this is not always the case: many high-cost headphones are quite unsuitable for the mobile recordist for reasons of lack of sound exclusion, weight or comfort. In fact some, although very comfortable to wear when standing or sitting, fall off when the wearer bends over to adjust the recording machine and may cause considerable damage. A recordist was recording the snorings of his sleeping dog when his heavyweight headphones fell off and hit the dog; the dog sprang up into the air; the recordist jumped back in alarm and smashed his television screen with the microphone. That was not all, because the cat was asleep on top of the television set at the time and when the tube imploded with an enormous bang the cat went wild and landed on top of the budgerigar's cage. What happened next—well, the chain of disasters went on.

Monitoring headphones should have a good frequency response (20Hz–20kHz) and, even more important, both earpieces should be of identical performance. The impedance must be correct for the headphone output jack on the machine. Often it is a case of taking the machine to a shop to see what sort of headphones it will drive: low-impedance (8Ω), or medium-impedance ($100\text{–}600\Omega$). The output volume should be sufficient to monitor in noisy surroundings. One criticism I have of the Nagra 4 is that the headphone monitoring level (maximum) is barely adequate, whereas the Sony TC510 gives more than adequate level.

To make critical judgement of recording quality and perspective, good isolation from environmental sounds is needed, and this must be checked with no modulation input to the headphones to hear how much outside sound penetrates the headphone ear-pads or transducers. Many headphones with large ear-pads have open backs (for extra bass response) which allow external sounds to enter via the diaphragm. The open-type headphone, which allows the wearer to hear sounds from outside as well as from the headphones, may be satisfactory for studio and home use—where total isolation from the rest of mankind is imprudent—but they are not suitable for fieldwork, because there will be confusion as to what is on the headphones

84

and what is outside.

Headphones should be comfortable to wear, as they may have to be worn for long periods in both hot and cold weather. The cord should not tangle easily: the coiled type is quite good in this respect (if it has not been ruined by being wound round the headphones for weeks and lost its springiness) because it allows a lot of movement and gives the recording machine some protection against sudden pulls on the cord. With non-coiled lead, damage can easily happen as a result of a twisted lead being snatched and accidentally dragging the recording machine to a lower resting-place.

The colour of headphones needs careful consideration. It is wise to avoid those colours which are prevalent in the areas most often worked in: to use green headphones in a field of new long grass is asking for trouble; black is inadvisable in a coalmine. A recordist working with a film crew at a construction site put his headphones somewhere and just could not find them. Two days later they were found hanging on the side of an earthmover— one of those yellow giants often seen trundling mountains of earth across the road at temporary construction-site crossings. The recordist thought he could not lose headphones of the colour he had chosen so when, pleased but somewhat abashed, he collected his yellow headphones he bought a small can of paint and painted one of the ear-pieces bright red. He has not mislaid his headphones since and finds the dual colours an advantage for stereo working.

5
Stereo

After listening to stereo for some time, monophonic sound from one loudspeaker sounds very dull and constricted—assuming that you have two good ears. The mono recordist advancing to stereo will find additional problems but the results will be rewarding.

Becoming attuned to stereo

Before beginning to experiment with stereo recording it is wise to become attuned to the medium, and that means listening to good stereo for a few hundred hours, spread over several weeks. This is to train the ears and the mind to accept stereo sound as the normal listening medium. Even the very best stereophonic sound will sound unbelievably constricted to those used to working with more advanced systems of sound reproduction: stereo sound is a far cry from natural sound, although it is much more natural than mono. Stereo is limited to sounds coming from a fixed angle, and the angle is determined by the relative positions of the listener and the loudspeaker units. Sight is limited to an angle which gives the viewer about 75° forward vision, and this seems to be the ideal angle for the listener to be positioned in relation to the reproducing loudspeakers in a stereo system. Stereo is not just sound from two fixed points but—what is more important—sounds which appear to emanate *from* and *between* two fixed points.

The degree of realism perceived by the listener depends on the degree of perfection in the recording and reproducing systems, and this includes microphones and their settings. With perfection achieved in all links the listener will be able to discern with pinpoint accuracy the position of every sound reproduced, and the sounds may be directly related to their original position in the field (synthetic or simulated recordings excepted). It cannot be emphasised enough that hour upon hour of good stereo listening is vital before starting to experiment with recording in stereo; otherwise, it is like trying to run before you can walk. Listening should be on a well-matched loudspeaker system set to subtend an angle of about 75° to the listener. When the ears and mind become well attuned to this 'hot-seat' position, stereo headphones should be used (if the experimenter intends to use headphones at all) and the listener should gradually learn to relate the positions of sounds heard on these to those heard on the loudspeakers.

Directional information

How do we detect the direction from which a sound originates? From the very early stages of childhood, noises cause the attention to be focused on the source of the sound. Babies just a few weeks old will turn their eyes to look for a visual indication of the origin of a sound. With continued practice and experience from the wealth of sound around it, a baby soon begins to correlate the two stimuli—sight and sound—and in this way the child's brain becomes programmed to connect the significance of time, phase and intensity differences received by the ear from sounds originating from different spatial positions. Children who have been blind from birth have received no visual stimulus to assist them in the location of sound sources; when asked to face the direction from which they believe a sound comes, many blind-from-birth children are up to 60° out in their assessment. It is a fallacy to say that the sense of hearing of a blind person is much more acute than that of a sighted person. Sighted people are perhaps less aware of the significance of the sounds around them, but they are better at assessing the direction and distance from which a sound originates; this is due to the continuous directive reprogramming which the brain experiences by way of visual stimuli.

The way in which the direction of sound sources is detected is very complex; each individual receives different information, owing to differences in head and ear shapes and in ear-to-ear distances. But each person receives the essential phase, time and intensity differences so necessary for the brain to compute the direction of a sound with pinpoint accuracy. Our ears (being on the sides of the head) are sensitive to direction in the horizontal plane only, and if the head is firmly fixed in position so that tilting is not possible, the height of a sound source cannot be comprehended. Height of sound is assessed by slightly tilting the head to allow the ears to be on relatively different horizontal planes. Also, whether we realise it or not, we usually direct one side of the head towards the sound source to assist location.

Loudspeaker stereo

In a stereo system it is vitally important that additional phase or time differences are not introduced in recording or replay. The information must be presented in an amplitude (volume) form only. The ears and brain can then compute correctly the direction of a sound reproduced on two loudspeakers, as if the listener were at the position of the microphones. The result is limited to an angle of some 75° horizontal and has no vertical information—however much one likes to imagine it has. When the amplitude of the two signals from the loudspeakers is equal and perfectly in phase, the sound will seem to originate from a point midway between the two loudspeakers. However, if the phase of one of the signals is reversed (negative or antiphase) the brain will be unable to agree on a location for the two signals, because they will not combine into something meaningful; the effect will be sheer torture as the listener's mental computer tries to reconcile conflicting signals.

In stereo drama it is permissible to use a small amount of negative phase, say 20°, for special effects—creaking doors, scratching on windows and many electronic sounds—to heighten the air of suspense, for these sounds will be outside the stereo sound-stage (see page 90) and difficult to locate, but not unpleasant when limited to very short durations. Sadly though, the mono listener must remain oblivious to the refinements which go into stereo productions.

Stereo, then, works correctly only when amplitude is the sole factor causing the brain to respond to the information given, so that when the loudspeaker on the left reproduces a signal identical to that on the right, but at a higher level, the resulting mixture appears to a listener in the 'hot-seat' position to emanate from a point somewhere between the left-hand loudspeaker and the midpoint.

Loudspeakers and phasing
Phase differences must be minimised through the entire system, and this is why it is important to ensure correct tape-head azimuth on recording machines. I have heard stereo systems (when invited to 'come round and hear my stereo') on which the enthusiastic owners proudly presented the material of their choice through out-of-phase loudspeakers. On occasions like these, a minute spent behind one of the loudspeakers usually results in exclamations like: 'That sounds super! What did you do?' I find it difficult to understand how anyone can sit and listen to out-of-phase loudspeakers without anguish: either they have come to accept their tortured offering as 'stereo' or they are undiscerning. I do not believe it to be the latter, because when the error is corrected they are usually the first to appreciate the difference.

Whatever system one uses, it is essential to ensure correct phasing. When mono sound is played through a pair of loudspeakers, the sound should appear to issue from directly between them (at least from the 'hot-seat' position). If there seems to be some confusion of sound image, try reversing the connections to *one* of the loudspeakers to see whether the output is positively or negatively phased. There will most certainly be a great difference (from the 'hot-seat' position) when the reversal is done. For those fortunate enough to possess a tape machine with marked output terminals, all that should be necessary is to ensure that the marked connections on amplifier and loudspeakers are identical. Loudspeakers which have been returned after repair are suspect, and a phase check should be made before installation.

A dc test on the main units can easily be done if the cone of each is visible. Take a 1½V battery and connect it across each loudspeaker in turn, making quite sure that the polarities are the same for each. Watch the direction of movement of the cones and mark the terminals which cause the cones to jump outwards when the positive battery terminal is connected. These will be the in-phase connections which should be connected to the positive output terminals on the amplifier. Make sure, also, that the interconnecting cables are correctly phased and then test the loudspeakers on signal. If a mono test still does not present a good central image (use the balance control) then either the main units have dissimilar frequency responses or the auxiliary units—tweeters and mid-range units—are incorrectly phased. Try covering completely all auxiliary units to determine where trouble may lie.

Hearing balance

In stereo, the balance of the ears is as important as the balance of monitoring loudspeakers, and if a clearly defined central image is not heard between two known-to-be-balanced loudspeakers, it may well be the individual's hearing which is suspect. One of my engineer colleagues has unbalanced hearing: the frequency response of one of his ears is vastly different from the response of the other, and one ear also suffers a large notch at around 1.8kHz. It was discovered, much to his mortification (my colleague having always maintained that he was unable to tell whether mono or stereo sound was being reproduced), when I conducted hearing tests on a number of volunteers. Perhaps I am fortunate in having ears balanced to within 1.5dB left to right from 20Hz to 14.8kHz (at the last test) and this is probably the reason why I get very well-defined stereo with pinpoint-sharp images.

The sound-stage

In the following section on microphone techniques for stereo, frequent reference will be made to the 'sound-stage'. A listener seated in the auditorium of a theatre is presented with an angle of sound determined by the working width of the stage and the stage-to-listener distance. Perspective (the depth of the stage) is also presented. The perspective ratio depends on the distance between listener to nearest performer and listener to furthermost performer, and on the acoustic properties of the stage area.

The acoustic information received by a listener in an auditorium can be relayed electrically to a listener seated between, and facing, two loudspeakers—and quite remote from the theatre. The distance between the loudspeakers, together with the angle subtended to the listener, represents the stage width and stage-to-listener distance. Thus, the listening-front—whence the sound appears to originate—becomes the sound-stage. A true image of the original stage sound can be conveyed only to listeners on a centre line between two loudspeakers, otherwise the image will side-slip to whichever side of centre the listener is biased. Some side-slip is also experienced, in a way, in the theatre auditorium by listeners when they move from a central position to a side position.

In Fig 19 imagine that the left and right sides of the proscenium arch of the stage represent the left-hand and right-hand loudspeakers, respectively, of the stereo listener at home. P1 and P2 are listening positions both in the auditorium and at home. The positions of the participants are W,X,Y and Z and the direct sound paths to the listening-points are marked across the

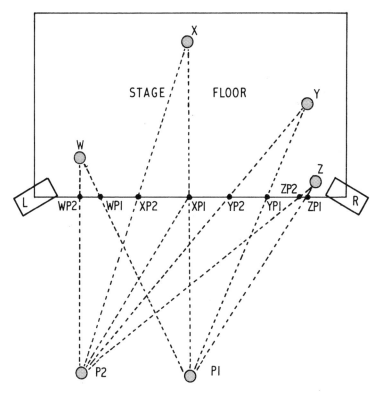

Fig 19 Theatre stage, showing the positional sound-slip experienced by an off-centre listener

stage front (also the sound-stage for the stereo listener).

It will be appreciated that position P2 has a smaller overall angle of direct sound than P1 and a difference in volume between W and Z (Z being about one and a half times the distance that W is from P2). Sound intensity diminishes in inverse proportion to the square of the distance; therefore at P2 there is an intensity difference of at least 2:1 in favour of W in direct sound. The auditorium listener at P2 experiences a slip in position of the stage sound only when the participants move upstage. (In the theatre the stage floor is usually raked forward to face the audience. 'Upstage' and 'downstage' thus refer to the back and front areas respectively.) When X moves from a downstage position to an upstage position the sound remains on the centre line for P1, but for P2 it slips sideways from the centre position XP1 to XP2. The stereo off-centre listener will experience side-slip on any material not hard left or hard right—see Fig 20(d).

Genuine and synthetic stereo

Sounds which come predominantly from one or other side loudspeaker cannot be classed stereo and they could even originate from locations and at times quite different from the main stereo material. Genuine stereo is time- and space-related. A great many recordings (particularly those of popular music), although labelled stereo, are not genuine stereo because they have been recorded using either time-unrelated material (multi-tracking, in which one or more recorded sounds are played back as further tracks are being recorded) or spatially unrelated material (different performers or groups of performers in acoustically different locations using headphones for time-relation purposes). In multi-track and similar recording it is usual to keep natural acoustic reverberation as low as possible because it might otherwise result in 'pools' of acoustic coloration round each 'panned' (electrically positioned) sound. The individual mono sounds are usually panned to achieve a good left/right balance—some hard left, some hard right and some across the middle. The whole is then usually augmented by some form of echo.

This type of recording is easy and enjoyable and there is always an air of anticipation in the proceedings, for exploration and discovery is the order of the day. Panned electronic stereo is not natural stereo but it does produce the illusion of width and space and, on the whole, the material contained in synthetic stereo recordings is not restricted by convention as regards positioning in the sound-stage. Further, it is produced to provide no more than pleasurable listening.

Genuine stereo and synthetic stereo can convey the same basic information, but where genuine stereo differs is in its ability to convey realism of movement and perspective. Genuine stereo is at its best when the natural environmental acoustics of recording locations are fully utilised—where the multiple reflections of the primary sounds are time-, phase- and space-related to the primary sound and to the environment. Perspective is thus conveyed not just as volume differences but as spatial differences too.

To stick a microphone in front of each performer in a small group of musicians in a non-reverberant acoustic, to record each on a separate track of a multi-track recorder and to think about balance and positional problems later, is fairly simple for the recordist. To achieve genuine stereo from a similar group of musicians in a live natural environment, and to get the balance of the performers and of the reverberant sound correct at the time of recording, may take a lot more thought and judgement.

Fig 20(a) shows two listening positions, P1 and P2, for a stereo system. P1 is the ideal position, which gives about 75° balanced listening angle and will still be referred to as the 'hot seat'. Fig 20(b) shows the perfect sound-stage to be received by the listener at position P1. The loudspeaker positions are indicated by solid and broken lines at 1 and 2, and 8 and 9 on the base grid of the sound-stage. The relative volume balance across the sound-stage is represented by the vertical height.

It can be seen from Fig 20(a) and (d) that the off-centre listener at P2 receives a very poor stereo balance. This is an imbalance similar to that which the hot-seat listener would experience if the balance control on the reproduction system were grossly offset. The off-centre listener also receives errors in phasing. These result from time-delay differences in the signals travelling from each loudspeaker.

93

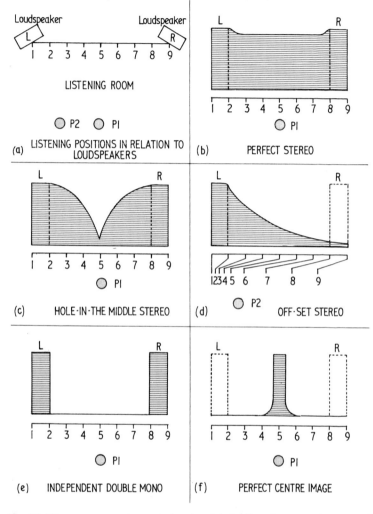

Fig 20 The stereo sound-stage, showing (a) the listening room layout and centre and off-centre listening positions (b) the perfect stereo sound stage (c) hole-in-the-middle stereo (d) off-set stereo (e) independent double mono (f) perfect centre-image

Microphones for stereo

Having established that only amplitude (volume) information is required for perfect stereo, some of the many ways of using microphones to produce the desired stereo effect will now be outlined.

In 1971, with my colleague John F. Burton (organiser of natural history sound recordings at the BBC), I embarked on a series of experiments aimed primarily at solving problems associated with wildlife sound recording. I was attempting to determine what could be done with microphones to obtain good stereo, and along the way I discovered some things which do *not* give good results. The 'hole-in-the-middle' effect is as common to beginners' experimental stereo as wrong notes are to a student of the Scottish bagpipes. 'Hole-in-the-middle' describes the condition where lots of sound emanates from both loudspeakers but very little seems to come from between them, as shown in Fig 20(c). Most of the sound should appear to come from between the loudspeakers and just a small amount directly from each. A hole in the middle of the sound-stage indicates that the microphones were either of a type unsuitable for the use for which they were used or that they were used incorrectly.

SPACED MICROPHONES

Because stereo listening is via two loudspeakers with space between them, it is often assumed that the actual recording of stereo requires the microphones to be spaced. Spaced microphones are used to record, say, operas, though they are used only as reinforcement to a main stereo system when required.

Consider two spaced microphones, L and R in front of a stage. Sound coming from the stage centre, 5A, B and C, arrives at each microphone at the same instant and at equal intensity and the result is stereo centre. When sound originates from the left side of the stage, at 1B and 2C, there will be phase and intensity differences. The intensity differences of about 4 : 1 will result in a predominantly left-hand position in the sound-stage. The result is the same when sound originates at 4A: there will be no apparent difference of position in the sound-stage. When sounds originate at 3A and 4A they will sound more to the left of the sound-stage than those originating at 1A and 2A. When the two

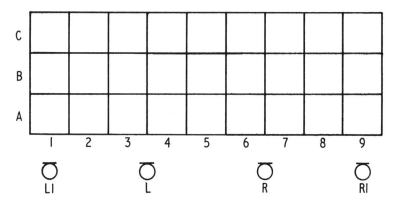

Fig 21 Spaced microphones positioned in front of a wide stage

wider-spaced microphones at L1 and R1 are employed, we can see that movement across the stage area is far better conveyed to the sound-stage. Unfortunately, the intensity level conveyed to the listener will be far greater from points near the microphones (1 and 9) than it will be from the downstage centre position (5). The effect in the listening sound-stage will be somewhere between that of Fig 20(c) and Fig 20(e). When all four microphones are employed and positioned correctly better results are achieved, but it is not a system which gives good results.

The mono-sound listener will receive a poor performance from spaced microphones, because phase errors cause cancellations at various frequencies and the sound quality will vary according to the position of the source. Two spaced microphones do not generally give worthwhile stereo: the wider they are spaced the more the result can be likened to independent double mono, the two signals becoming more and more unrelated, as in Fig 20(e).

CARDIOID MICROPHONES

When two omnidirectional microphones are positioned close together the result will be integrated double mono, giving a perfect central mono image in the sound-stage. The same is true of any two (or more) microphones of the same type, if they are very close together and facing the same direction. No phase errors can occur in this condition and this is ideal. It follows,

96

therefore, that two cardioid microphones will produce perfect double mono when their heads are very close together and facing the same direction. It is better that the heads are in vertical alignment, for then there can be no time (or phase) differences from sounds arriving from different directions in the horizontal plane—the plane in which the stereo listener hears movement.

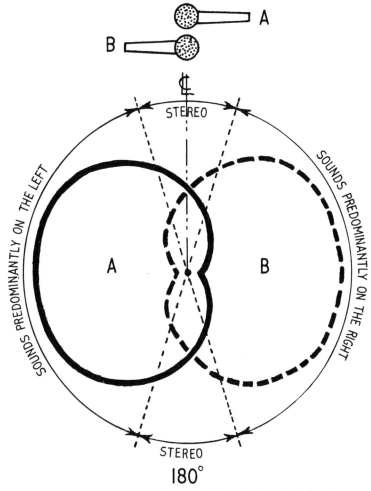

Fig 22 Cardioid microphones positioned back-to-back (180°)

This vertical alignment is commonly called coincident. However, by definition, coincident means 'in exactly the same space or time', and as this cannot be achieved with two or more objects of finite size, it is more correct to refer to the positioning as near-coincident. This may seem a pedantic point, but we shall see that distances as small as 5cm between the microphone heads in the horizontal plane can have most untoward effects on the frequency response of signals combined for mono listening.

If two in-line cardioid microphones produce good centre-image double mono—as though only one were being used to supply both loudspeakers—then what will be the result of turning one of the microphones through 180°, still keeping both heads in vertical alignment? What happens in the vertical direction is not particularly important and will be ignored for the moment; we shall consider only the horizontal performance.

In Fig 22 the polar response pattern of microphone A is in the opposite direction to that of microphone B and the phase of signals is positive at all angles. The combined polar response is therefore spherical (and better than many single omnidirectional microphones), because at 90° all the way round the centre axis of each microphone the response is 6dB lower than it is on-axis; when the two responses are combined, the result is 0dB level—as on-axis. Any sound arriving at 90° to the microphones—sideways on—will be reproduced in the centre of the sound-stage. A cardioid microphone has a back-to-front ratio of 20dB, and those sounds which arrive on-axis to one microphone will arrive in the dead zone of the other and vice-versa.

For back-to-back cardioids the zone of acceptable stereo—the portion where movement and positions are clearly conveyed—will be found to be rather narrow (about 35°), as indicated in Fig 22. All sounds from outside this limited angle will reproduce predominantly on the left and/or right and will give a pronounced hole in the middle, as well as 'blocks' of mono sound at each loudspeaker. In close-speech use, movement becomes exaggerated and mouths sound wider than life size (if this can be imagined!).

There is an angle between 0° and 180° which suits two cardioid microphones. The angle to be used will depend on where the bulk of the sounds in the surrounding area originate. Where the sounds are from a limited angle, say 120°, the angle between the axes of the microphones can be about 90°. Where sounds originate from 360°, as in most outside work, a more acceptable result will be achieved when the inter-axis angle is reduced to 50°. The exact angle for any type of work will have to be found by trial and error and may vary with different microphones. The background sound will also have a great bearing on the angle to be used because, although for the required forward sounds the angle might be 90°, this could be an excessive angle for the background atmosphere of any environment where recording takes place. When this is the case both loudspeakers will reproduce the atmosphere in block mono images—too much on the left, too much on the right.

Fig 23 shows the conditions of combined polar responses for a pair of coincident cardioid microphones at 90° and at 50°. Notice that neither arrangement offers much in the way of a dead zone (angle of low sensitivity) and this can be a big disadvantage. Cardioid microphones are not really suitable for recording small-angle subjects in areas of high ambient sound, particularly if the ambient sounds detract from, rather than enhance, the primary sound. On the credit side phase errors should not occur, whatever the angle, because cardioids have no negative phase zone.

The importance of coincident positioning for two microphones used as a stereo pair has been emphasised. Strange things happen to the frequency response when the heads are separated, as I discovered when wildlife recording. I use a pair of AKG C451 cardioid capacitor microphones for recording close to nesting birds and for other close-range activities. Knuckle-joint devices are used to angle the capsules—the active head units—at 90° to the bodies of the microphones. The microphones are mounted vertically and side by side in an all-enveloping, specially constructed windshield.

At one time I used the capsules pointing outwards, allowing a space of roughly 5cm between the active part of each head capsule. This arrangement was used to record the calls of a common sandpiper feeding at the edge of a shallow pool. The microphone unit was concealed in a clump of marsh grass and

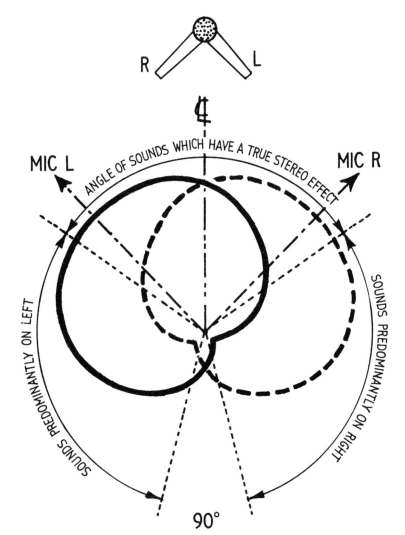

Fig 23 Coincident cardioid microphones at 90° and 50° indicating the
equivalent sound-stage results

the bird walked to and fro along the muddy edge of the pool, very
close to the microphones. This resulted in a close-perspective
recording of the bird's piping as it called or answered more
distant sandpipers—altogether a very satisfying recording of a
perky little bird. However, in the processing stage (copying to

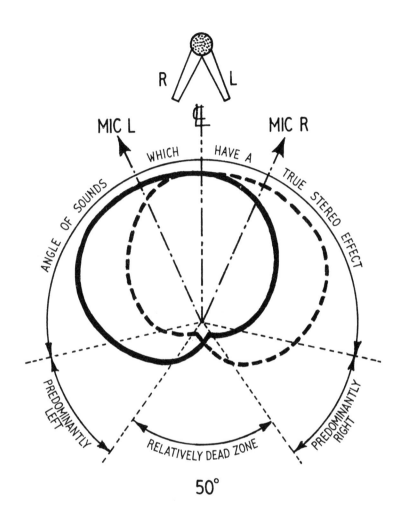

MIC L   MIC R

R   L

WHICH   HAVE A

ANGLE OF SOUNDS

TRUE STEREO EFFECT

PREDOMINANTLY LEFT

PREDOMINANTLY RIGHT

RELATIVELY DEAD ZONE

50°

master tape) it was discovered that when the stereo recording was played in mono (both stereo tracks combined) the piping disappeared when the bird would have been at 60° or thereabouts from the front of the microphone assembly.

The reason for this peculiar result is quite understandable

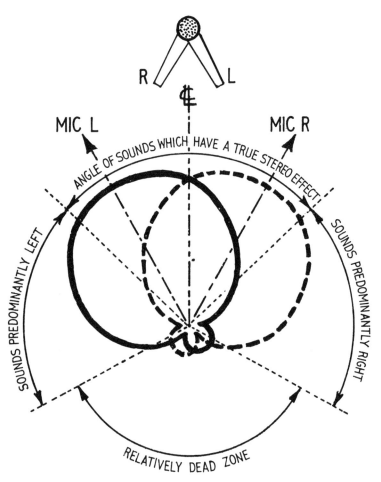

Fig 24 Coincident super-cardioid microphones at 60°

when one thinks about it. Complete cancellations occur when the spacing is a half-wavelength of the received frequency. In this case the sandpiper's call was of a frequency near 3kHz, and half a wavelength of 3kHz is 5.5cm—roughly the distance between the two microphone capsules—and this was responsible for the phase cancellations when the recording was played in mono. The outward-facing angle was then dispensed with in favour of overlapping inward-facing coincident capsules.

The sandpiper recording was not a total disaster, you may be pleased to hear, because it was possible to correct the phase error by carefully varying the azimuth of the tape head on the replay machine while copying the recording. It is not a practice which I would advocate because, owing to the bird's continuously varying position, several runs had to be made before an acceptable result was achieved—quite a boring exercise!

SUPER-CARDIOID MICROPHONES

A super-cardioid polar response is more directional than a cardioid; where the rear lobe, if any, is very small, super-cardioid microphones are better for stereo than cardioid microphones. A super-cardioid response is commonly −6dB (half level) at 70° from front axis, whereas with a cardioid the −6dB point commonly occurs at 90° from front axis.

It can be seen from Fig 24 that, when the inter-axis angle is 60° the zone of acceptable stereo will be around 95°. Note that compared with cardioids (Fig 23) super-cardioids provide a more worthwhile dead zone. Back-to-back super-cardioids provide a low sensitivity to the stereo front position and used thus they would not provide acceptable results. The angle most suitable for a pair of super-cardioids will depend, as for cardioids, on the precise response of the microphones and on the nature of the recording for which they are to be used.

HYPER-CARDIOID MICROPHONES

Hyper-cardioid microphones are suitable for stereo; Fig 25 shows the condition quite typical of a pair at an inter-axis angle of 65°. As will be apparent, the more a microphone becomes bidirectional the less becomes any dead zone of a pair used for stereo. The frontal stereo angle is fairly wide but care needs to be taken when use is made of the extremes, because the phase of the rear lobe of hyper-cardioid microphones is opposite to the front. Stereo listeners experience unpleasant results when sounds originate in areas of opposing phase; this out-of-phase area is a zone where the sensitivities of the front and rear are at a ratio of roughly 6 : 1 or less. The worst effect occurs when the ratio of positive to negative phase is 1 : 1 (which is also the condition which gives total cancellation in combined mono).

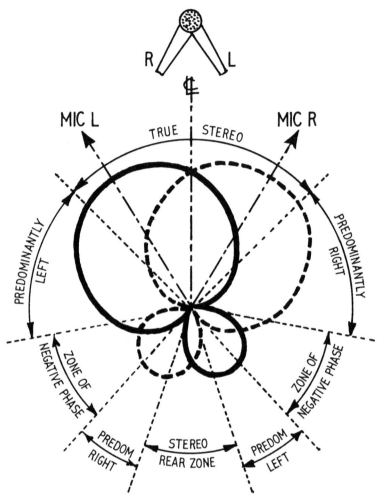

Fig 25 Coincident hyper-cardioid microphones at 65°

The rear lobe area of a hyper-cardioid pair may be used for stereo if the polar response of both rear lobes is uniform. The rear sector will have about half the sensitivity and half the usable angle of the front sector. This can be very usefully employed. For example: an actor is required to shout in a large reverberant acoustic to achieve the effect of a speaker addressing a crowd. The speaker at the rear sector of a hyper-cardioid pair of microphones will be facing the crowd, which will be on the main stereo sector, and he will thus react in a natural manner with a raised voice. The crowd should be positioned widespread and at differing distances from the front of the microphones. The result should then be a speaker whose voice is clearly heard above a crowd which has good perspective depth.

A point which has been mentioned already, and is perhaps worth stressing again, is the apparent confusion in some data concerning polar responses. In the data of one manufacturer a particular response may be referred to as hyper-cardioid, whereas a different manufacturer refers to an almost identical polar response as being super-cardioid. The point is that it is important to become acquainted with the precise polar characteristics of a microphone in order correctly to set the angles appropriate to the work to be undertaken. The results in practice may otherwise be rather different from those anticipated.

I find it very useful and informative to carry out simple practical tests on microphones, and these tests are particularly valuable in helping to determine the performance of stereo systems. I have done, and still do, these practical tests out in the middle of grassland in open country where reflections are, for all practical purposes, zero.

The microphones to be tested are set up on a stand about 2m high in an area of not too tall grasses. A piece of rope about 3m long is centred at the point where the microphone stand will be and, while holding the free end of the rope at a set distance from my body, I tread round making a well-marked, complete circle. Further marks are made at 30° intervals round the circle. The microphones are placed in the centre and directed towards the 0° marker, and are then connected to a recording machine. I usually walk round the marked circle, facing the microphones as I do so, and in as constant a voice as possible call out the angles as they are passed. On listening to the results the areas which require a more detailed examination will be obvious and further tests are then confined to those areas. If constant-frequency tones are available from a portable generator they can be used to do very detailed polar response tests of high accuracy.

The normal method of testing a microphone is to place it in an anechoic chamber with a sound source at 1m distance; the microphone is rotated about an axis while pen-recording the result, using a standard calibrated microphone for comparison. Anechoic chambers are specially constructed rooms with sound-absorbent materials over their entire surface area, including the floor and door, and are very 'dead'. However, they are none too freely available, whereas open grasslands are.

Stereo tests in open grassland are especially informative, as the merits and disadvantages of a system become apparent immediately upon replay of the test material. Different inter-axis angles will result in differences in volume of sound as the source traverses the sound-stage. The areas where there is no apparent movement in the sound-stage to match an actual movement in the field (block mono areas) will also be heard. This test method may seem laughable, but it really does show the practical points which are required to be known about microphones. It is largely from tests carried out in the manner described that many of the findings in this book are based.

On one occasion a passing cyclist stopped to look on at my seemingly odd ritual. For some while he just watched me walking round in circles talking to the centre pole, then after scratching his head he threw up his hands as though to say 'There are more out than in' and went hastily on his way with an apprehensive backward glance.

FIGURE-OF-EIGHT MICROPHONES

Crossed figure-of-eight microphones should be mounted as nearly coincident as possible. In the electrostatic variable-polar-response stereo microphone one capsule is directly above the other and the uppermost capsule is rotatable through $\pm 90°$. The angle between the axes of figure-of-eight microphones will depend on the amount of sound originating from the sides (the out-of-phase zones) and also on the polar response of the microphones at high frequencies. Some figure-of-eight microphones have rather narrow angles of acceptance at hf and, when angled at 90°, they produce a noticeable hole in the middle of the sound-stage at hf. To counteract this the angle may need to be reduced to something nearer 60°.

With figure-of-eight assemblies, the front and rear sectors are, in a stereo sense, interchangeable or reversible. In other

106

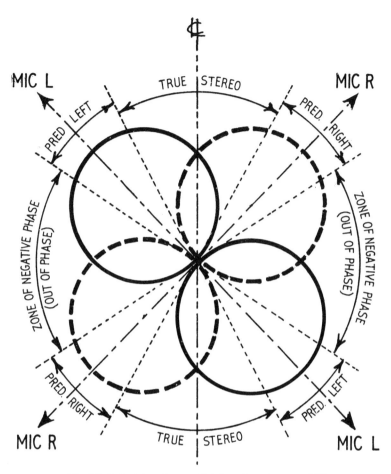

Fig 26 Coincident figure-of-eight microphones at 90°

words, no difference should be noticed when the system is turned round the other way (180°), because the front and rear sensitivities are equal. The phase of the rear, however, is in opposition to the front and signals which arrive from the sides will be received by the positive face of one microphone and by the negative face of the other. We now know well enough that out-of-phase signals create very unpleasant effects for stereo listeners and should be avoided.

Unfortunately, the very efficient cancellation properties of the figure-of-eight microphone—the plane of the ribbon rejection—cannot be utilised in stereo because the dead zone of one microphone is usually positioned to coincide with the 0° axis (front) of the other. A high rejection is provided in the vertical direction but this can be used only when an unwanted sound is either above or below microphones used in the normal vertical way. To take advantage of the rejection zone to cancel out a particular horizontal sound, the microphones would need to be used in a horizontal plane, under or over the desired sounds.

GUN MICROPHONES

When gun microphones are used for stereo, the phase of the two signals will nearly always be different and will vary as the frequency and angle of the originating sound varies. Fig 27 shows two gun microphones at 28°, an angle which proved to be the optimum for the microphones I used, the Sennheiser MKH815.

When sounds arrive at each microphone at the same instant, the phases of the output signals are the same. This is the condition at the centre line between the microphones, but as the angle of sound deviates from the centre line so will the time of arrival of the sound at each microphone differ. The high frequencies will, in fact, change from positive to negative phase several times when their direction changes from the centre line to the axis of one of the microphones. This will not cause too many problems in material of multiple upper-frequency content, although the signal wavelengths will always be several phases out on account of distance between the microphones (5.5cm is the cancellation distance at 3kHz), because the ear is not too critical of phases at high frequency. The low- and mid-frequency ranges are the power ranges; the higher frequencies do not add power to the sound—it is not their function. Their function is to add clarity and intelligibility to the power range and this they will do even if the phase is a few wavelengths out of step. The ear becomes more critical of signal phase as the frequency decreases from about 1kHz. The distance between a listener's ears is roughly 33cm, which corresponds to a full wavelength at 1kHz; this is perhaps the reason for the increasing sensitivity of the ear to phases below 1kHz.

The distance between the fronts of a pair of gun microphones

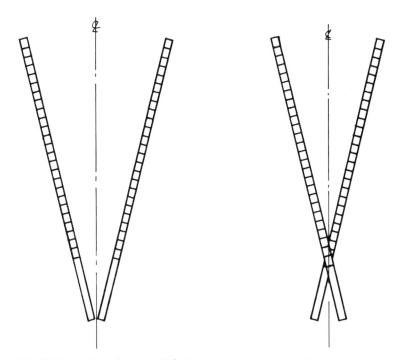

Fig 27 Gun microphones at 28° showing how one pair (right) is crossed to minimise phase errors

is about 30cm and the phase of signals originating off the centre line will be modified. The wider the microphones are spaced the greater the problems become. I have experimented a great deal with gun microphones as stereo pairs and I find the insurmountable problem of phase error a serious limitation to their use. The least overall errors of phase were obtained when the tubes were crossed one over the other at 28° and at a point about 5cm in front of the last sound-slot. The phase problem is particularly serious at the extremes of the pick-up angle, which is 70° total, although this is not fully apparent until the results are heard in mono. Then it will be most obvious that the quality of sound changes with the position of the source. For example, the mono replay of a person talking while walking from one side of the acceptance angle of the stereo microphone system to the other side will be heard to vary considerably in tonal quality.

This effect can be used to advantage, particularly in rather dramatic acoustic locations, such as tunnels, underground chambers or other situations where acoustic quality is important. By using the phasing effect of a pair of gun microphones, the acoustic quality of a performer travelling across the acceptance angle (the field sound-stage) may be heightened, thus conveying a situation more realistically or more dramatically than could be conveyed by the acoustics alone. Wildlife sound recording is an area where small variations in quality are generally acceptable, and a pair of gun microphones will do justice to a wide range of natural sounds.

The parabolic reflector is the only device capable of pulling in sounds from considerable distances and this very narrow-angle device can be adapted to produce a reasonably acceptable stereo performance. It will be discussed in Chapter 9.

Super-stereo

For any pair of microphones there is no pre-set angle which will suit all occasions. Each situation may demand a different angle to maximise its stereo potential. Cardioids at 50° axes will effectively condense about 300° of field sound to produce good stereo at the 70° sound-stage. But in locations where all the sound comes from a limited angle of, say, 120° the microphone axes may be increased to about 90° for cardioids and 75° for hyper-cardioids. As the angle of field sound *increases* the angle between the microphone axes should *decrease*. To cause field sounds to sound closer together in the sound-stage the microphone inter-axis angle is decreased. This is obvious when one considers that at 0° inter-axis angle there will be no stereo at all, only centre mono, and that maximum angles produce maximum stereo separations and often produce independent block mono.

Many manufacturers produce single-unit stereo microphones in which the angle is pre-set and non-adjustable. These are the one-point microphones intended for domestic and semi-professional use; the elements are most usually cardioid or uni-directional electret and angled at 120°. The one-point microphone is, therefore, suitable for recording field-sound angles not greater than about 100°. The elements are not

110

necessarily coincident: they are more likely to be mounted side by side, with about 5cm space between them. The one-point microphone is usually neat and convenient to use and is capable of giving good results in a limited range of applications.

To provide the listener in the stereo 'hot-seat' position with a good stereo image lacking any suggestion of a hole in the middle or any noticeable quantity of fixed block mono to the left or right, and at the same time present the side-seat listener with a good stereo effect, can be difficult. With two microphones it is possible to satisfy fully only one listening position, at the expense of another.

It is possible, however, to satisfy both positions: to produce a good stereo effect over a wide area—and without detriment to the 'hot-seat' listening position. It can be achieved by using a combination of the maximum separation given by two widely angled microphones, together with the minimal separation given by two narrowly angled microphones. The system was evolved during the course of the many wildlife recording experiments which I devised, and uses four hyper-cardioid microphones. Hyper-cardioids have very good dead zones at between 125° and 140° from front axis. When two are placed together with about 115° between their axes, a high degree of separation is achieved for two sectors of sound lying between 50° and 122° on either side of the centre line between the microphones. The sounds from these two sectors will be reproduced only by the appropriate loudspeakers and will not slip should the listener move off-centre. These two 115° angled microphones I call left and right outriders. They produce the very firm sides to the sound-stage, but on their own they produce very poor stereo, with a more than just noticeable hole in the middle.

A second pair of hyper-cardioid microphones are positioned with the first pair but with only 25° between their axes. These produce a narrow-width (low-separation) stereo image, but stereo nevertheless, with a well-filled centre. These I call fillers, and when the fillers are harmoniously married to the outriders, super-stereo is born! The left filler and left outrider are electrically interconnected—with careful attention to phase. Similarly, the right filler and outrider are interconnected. The paired left and right outputs are then taken to the inputs of a stereo microphone amplifier; they should produce 6dB less hiss as a result of the combined impedance being lower, or the output voltage being higher (not applicable to capacitor microphones), depending on whether series or parallel connection is used. The arrangement results in good positional information over a wider listening area and also provides the extreme side seat positions with a full-width sound-stage, though obviously the linearity will differ from the field-sound.

Results almost as good as those given by the four-microphone system can be obtained by using only three microphones.

In Fig 28, microphone A is stereo left and microphone B is stereo right. Microphone C is panned electrically to the centre of the sound-stage, ie the signal is fed equally to left and right. The centre microphone is used to control the apparent angle and image positioning of the sounds on a set—the sound equivalent of a zoom lens, one might say. By adjusting the balance of the centre microphone the sound in a large stage set of 180° or more

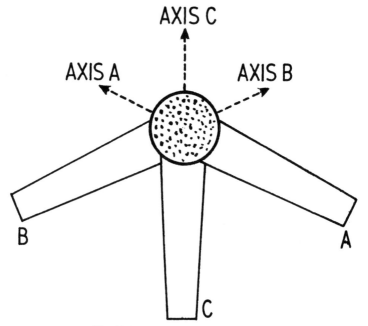

Fig 28 The three-microphone system

can be condensed to a 70° sound-stage with very good image positioning and good linearity.

Cardioid, super-cardioid and hyper-cardioid responses can be used in the three-microphone system. Even gun microphones can be used in this way and an angle of 20–25° between each would probably produce a good result.

Earlier it was mentioned that the dead zone of certain microphones can be fully utilised in mono for noise rejection, but in stereo these rejection properties are less useful. This is true of any pair of microphones, but if one is prepared to employ additional microphones and use them as cancellation devices to overcome the lack of a good rear dead zone—for example, by using a fourth cardioid microphone positioned coincidentally with the three cardioid microphones in a three-microphone system—it is possible to produce a stereo system with a very good rear rejection zone. The fourth cardioid is positioned with its front axis facing to the rear, but its output is phase-reversed and fed to left and right channels equally, via a mixer. By carefully adjusting the gain balance between the outriders and the fourth (phase-reversed) microphone, a substantial rejection zone is produced at the rear of the assembly without affecting the stereo front zone.

A hyper-cardioid already has a phase-reversed rear pick-up zone and this can be employed in a similar manner to produce a good dead zone, thus avoiding the need for a fourth microphone. Instead of using all cardioids in the three-microphone system a hyper-cardioid can be employed for the centre microphone, with its output adjusted to produce the desired dead zone. There will be no need to phase-reverse the output of the centre hyper-cardioid microphone; in fact this should not be done, because it would then produce a dead zone in the centre of the stereo front zone, and the already phase-reversed rear lobe of the hyper-cardioid microphone would become additive to, rather than subtractive from, the rear pick-up of the outriders. The flexibility of width control would, of course, be lost in generating a good dead zone but, by careful adjustment of angles of microphone axes and of the balance of the centre microphone gain, a superior wide-angled stereo system with a good rear rejection zone may be achieved.

113

Stereo balance

Balance in the sound-stage is an important aspect in stereo recording and the utmost care should be exercised to preserve it. The stereo listener needs not only the desired sounds to be well balanced, but also any accompanying background. Extraneous sounds should not irritate the listener by being biased to one side of the sound-stage. If offending noises cannot be eliminated they should be balanced to give the least possible degradation to the sound-stage.

To appreciate the effect of balance on the listener, it is worthwhile deliberately offsetting the balance control on your listening-system amplifier. Using a good-quality source of stereo sound, listen for a while with the balance normal (centre), then swing the sound left and listen a while longer, and then swing the sound to the right and listen yet again. Sit in the 'hot-seat' position throughout. After a while close both eyes and adjust the balance control to obtain a nicely balanced sound-stage—only one position of the control will achieve correct balance. If, subsequently, a recording with a background biased to one side is played on the correctly balanced sound-stage, the listener will immediately feel inclined to readjust the balance, irrespective of the balance of the primary material. The more constant the noise the more is the need to balance it. In sound drama a clock ticking on one side is perfectly satisfactory, while with boats at sea in rough weather, to bias the rough sea and wind noise to one side would ruin the dramatic effect as far as stereo is concerned, and in all probability it would cause listeners to reach for their balance controls.

6
Recording Talks and Interviews

The recording of talks and interviews is one of the most creative fields. Interviews pose two challenges—as a sound recordist and as an interviewer. There is always plenty of material available. Some people seem always to be talking, while others are hard to get talking. In club bars worldwide, opinions on a limitless range of subjects and topics are openly expounded to anyone willing to lend an ear. The recordist can obtain, candidly, fascinating information, dialects, jokes and a wealth of amusing anecdotes, simply by being there. Of course, the concealment of microphones and recorder is most important in such situations.

I have recorded in stereo street festivities, games and sports groups, chatter in bars, buses and trains, concert-goers and market-place ambience without anyone realising it, simply by carrying a bag containing microphones and recording machine. For crowded locations the bag can be carried on the shoulder to gain height, and the microphones can be concealed under a thin piece of cloth or a scarf. Microphones can be concealed in hats, caps and wigs or even the recordist's own hair, in imitation books and parcels, and in the handles of walking-sticks or umbrellas—the only limit is the inventiveness of the recordist.

A great deal of folklore and local history is on tap from the lips of old-timers. Some react strangely to the presence of a microphone and so disguise may reap better results; on the other hand the presence of a microphone may cause others to speak more clearly.

When your recordings are 'in the can' the interesting stage of editing and assembly follows and from it you may produce your own sound programmes and documentary features—something to be proud of, and all of it original. Before you go rushing out to find an old-timer or a schoolgirl poet, however, let us discuss the handling of microphones.

115

Holding the microphone

Recording talks and interviews seems, at first, to be as easy as holding a microphone in front of whoever is to be recorded. It *can* be that simple if the right microphone is chosen and held in a way which avoids handling noises. These noises vary from low-frequency rumbles to high-frequency brushing noises. Many microphones have a metallic surface; should the fingers be allowed to brush this surface a high-frequency noise akin to that of brushing paper is produced—the lighter the body construction of the microphone the greater is the likelihood of producing the noise. Even light friction between the fingers of the hand holding the microphone can cause noise—place your hand to your ear and lightly rub your fingers together to hear the type of noise I mean. Experiment also with moving the fingers over various materials held close to the ear. It will be noticed that matt surfaces produce the most hf brushing noise and polished surfaces produce the highest 'grip' noise. The more moist the fingers the greater the noise on polished surfaces: the fingers adhere to the surface and when contact is released they come away with a sticky, sucking noise. These handling noises are transmitted direct to the microphone which, in effect, is the ear of the listener.

The output lead is another source of noise: it can twist and turn and touch against things, and in so doing it may transmit mechanical vibrations to the microphone. Rapid movements in the air (or in the case of many capacitor microphones, only slight movements) can cause a microphone's diaphragm to exceed the maximum permissible excursions and a distorted blasting sound is produced, even when wind-muffs are used.

If a microphone is to be hand-held for interviews, etc, it should be reasonably free from the noises described. The output lead should not be allowed to dangle from the end of the microphone, but should be doubled back and held against the body of the microphone, with the exit between the first and second fingers. In this way, any shock noises carried along the cable are damped by the fingers. A firm full-hand grip on the body of the microphone will help to minimise hand noise, and microphones of reasonably large diameter will be found easier to hold, although I have seen a thin microphone used with a handker-

116

chief wound round it to increase its diameter.

Avoid sudden movements with microphones, especially with zero-frequency capacitor types. If quite large movements become necessary, as when interviewing several people simultaneously, leave the moves of the microphone isolated, if possible, and ask the questions and receive the replies only when the microphone is stationary. The gaps containing the shattering noises produced by very quick moves can later be edited out. If, as is common, the answers begin with 'Well...', that 'well' may be a safeguard, because it probably happens just as the microphone is coming to rest after a quick move. Therefore 'well' has movement noises over it, and both the 'well' and these noises are removable in one edit.

A microphone should not be placed directly in line with the mouth: it will result in blasting; 'p'-blasting is a term commonly used to describe the resulting noise when words containing the letter 'p' are spoken. A blast of air is projected from the lips and when this hits the microphone a loud explosive noise is imprinted on the recording. The breath blast can easily be observed if a small quantity of talcum powder is applied to the end of a finger and held in various positions close to the mouth while speaking. When the position of the applied powder is directly in line with the mouth, any word containing a letter 'p' will immediately disperse the powder, thereby proving that a blast of air is produced.

If the powder test is repeated, it will be found that the best position is to the side of the mouth. A position just below the lower lip will be free from 'p'-blasts but will pick up air blasts from the nose. Special attention should therefore be given to the positioning of microphones if 'p'-blasting is to be avoided.

At public functions where a loudspeaker system is used in an attempt to enhance the clarity of the spoken word, it is all too commonly used incorrectly: instead of enhancing clarity, it results in a loud but muffled sound interrupted by frequent loud popping noises. The systems installed in public establishments are generally capable of producing reasonable results when used correctly. The difficulty comes in training those who may be required to use these systems, and this training by no means stops at the microphone. However, the microphone is a good starting-point.

117

Some years ago there was an artiste who would grab the microphone from its stand and blow hard into it before addressing the audience. Anticipating this one day, one of the sound engineers reverse-plugged the microphone (a robust dynamic one) to a separate talk-back microphone and amplifier. The artiste grabbed his microphone as usual and blew hard into it, whereupon the microphone reacted most strangely. It screamed back at him 'Stop it, stop it. Stop blowing me, I'm not a candle.' It cured him of the habit.

If a stand microphone is to be used at a function, a lectern or similar object should also be used. Some people seem to continually fiddle with any microphone they address, so causing all manner of extraneous noises. If a lectern is placed in front of the microphone, most speakers seem content to wrestle with the lectern instead, thus leaving the microphone free to perform its duty of sound reinforcement. The microphone should be positioned slightly to one side, so that when the speaker is facing the lectern the microphone is neither in the line of sight of, nor in the line of breath blasts from, the performer. Where a non-sprung stand is employed, it is well worth going to the trouble of placing it on rubber or felt pads to provide isolation from floor vibrations.

Probably the best results are achieved by using tie-clip personal microphones for such functions, but even these are not free from noise should the users pull their clothes about. Tie-clip or lapel microphones do not restrict a performer's movements as do fixed or hand-held microphones and there is also less self-consciousness on the part of the performer. When a radio microphone is provided, a performer can often forget that a microphone is attached.

The amplifier delivering the sound reinforcement should be carefully set to exploit the system fully. It is tiring to listen to loud, 'bassy' speech when what is required is enhancement of intelligibility. Best results are achieved by reducing the bass substantially (which, incidentally, will also help to reduce 'p'-blasting and floor bumps), increasing the middle-range frequencies between about 500Hz and 5kHz, and tailoring off the high frequencies. The volume need not then be very high to achieve clarity of speech. A recording feed taken from a performer's microphone should be taken at a point prior to any frequency corrections on the public-address amplifier.

Feedback or howl-round will occur when the volume level from the system equals that required at the microphone to produce that same volume level. The more directional a microphone and the further it is from the loudspeakers, the less is the chance of a howl-round; super-cardioid microphones will perform more satisfactorily than omnidirectional microphones. In professional public-address systems, howl-round is limited by various devices. Some shift the frequency of the amplified sounds by a few hertz to break the regenerative cycle, while others delay the sounds by a few milliseconds to ensure that the repetition rate cannot approach that which may regenerate.

Time delays can be produced by recording machines and by electronic devices. Large multi-loudspeaker systems, such as those installed in cathedrals, use multiple time delays to overcome multiple echoes. In a cathedral each loudspeaker is sent a sound-signal differing in its time of arrival; this results in a wavefront all the way down the centre aisle which is identical to that produced by an unaided voice at the microphone, but of course in an amplified form.

Choice of microphone

The ambient noise level will have a significant influence upon the choice of microphone for interview work. Where atmospheric sounds of the surroundings can add meaningful embellishment to the spoken word an omnidirectional microphone may be chosen, but where the ambient sound becomes an uncomplementary high-level noise an omnidirectional microphone may produce poor results. It is better in such circumstances to seek a quieter location. When this is not possible, consideration should be given to the use of a microphone which is likely to minimise the pick-up of disturbing noises. A figure-of-eight microphone held horizontally will cancel surrounding horizontal-plane sounds to an amazing degree, enabling interviews and commentaries to be undertaken in very noisy places, eg beside motor-racing circuits, heavy seas or pudding-mixing machines!

I shall now deal with some aspects of talk and discussion

recordings in various acoustics. In an ideal 'talks studio' acoustic (where reverberation of sound is very low) a group of speakers can gather for discussion round a hexagonal, octagonal or round table, with an omnidirectional microphone at its centre, and freely discuss among themselves without turning away from the microphone and without sounding as though they were talking in a bathroom. Reverberation in a room results in loss of intelligibility, and many people find echoey rooms very tiring psychologically. The reason for installing public-address loudspeakers in a large room is to provide localised close-perspective sound in areas where the direct sound is too coloured with reverberation to be intelligible. The more reverberant a room, the more directional should be any microphone used in it.

Setting up for recording in mono

A standard and much used studio set-up for two-person recording is as follows: the two contributors are positioned facing each other across a rectangular acoustic table just over a metre wide, and a figure-of-eight microphone is placed between them. Level balance is achieved by adjusting the relative distances between each contributor and the microphone—the microphone being moved rather than the contributors. The microphone is placed slightly to one side of the contributors' eye-lines, partly for their convenience but mainly because it is less likely to receive their breath blasts in that position.

An acoustic table is one with an acoustically transparent top. The usual construction for the top is a sheet of perforated metal (or plastic) fixed over a wooden framework and covered with cloth or some equally sound-porous fabric. The principle is that the table, while presenting an object round which contributors can sit, in no way modifies sound-waves, because it is virtually acoustically transparent. However, most contributors place some paperwork (notes or script) before them, thus presenting an acoustically reflective surface (at high frequencies) which partially nullifies the non-reflective characteristic. Also, because the table is acoustically transparent, any noises from under it, such as hand movements on knees, chair creaks and other things that happen under tables, are in a direct line with the microphone and are unattenuated by the table.

An acoustic table is built on a good theoretical principle but it does not, in my view, work too well in practice. I have found that a normal solid-top table covered with eight thicknesses of ordinary woollen blanket is far superior to an acoustic table. Blankets provide a comfortable soft edge and top to a table; they are also acoustically absorbent and available in a variety of colours and patterns. A solid top provides a good sound barrier against under-table noises.

The across-the-table figure-of-eight microphone arrangement can be used for a group of three or four contributors although, owing to the none too wide angles of pick-up of the microphone, the contributors at the sides of the table will have to sit fairly close together to avoid sounding off-mic (ie sounding as though at some distance from the microphone). The main problem with an arrangement like this is that, once the microphone has been placed in a position which achieves good balance, it is more difficult to control the balance should one of the contributors during the course of the discussion sit more forward or backward than the original balanced position.

In a three- or four-contributor group one person may act as the controlling chairman of the discussion and may wish to sit at the 'head' of a rectangular table. In a one-plus-two discussion the two contributors are positioned facing across the table and just over a metre away from the 'chair' end. A figure-of-eight microphone is placed between them—not on their eye-lines but about 30cm further towards the chairman. This is the best position for the microphone because, whether addressing the chairman or one another, the contributors remain on-mic. The chairman will be in a zero pick-up zone of the figure-of-eight microphone and a separate microphone with a hyper-cardioid response should be provided for his benefit. The positioning of this microphone is quite critical because a hyper-cardioid has good dead zones at about 130° and 230° from front axis; these dead zones should be directed towards the other two participants. In this way there will be a minimum of reaction between the two microphones and independent control will be possible.

When one-plus-three discussions are to be recorded in mono, two contributors should be positioned on one side of the table (side A) but not too near to one another, the chairman at the head (C) of the table and the remaining contributor seated just over half-way down the other side (B). Again, a figure-of-eight microphone is used for the two contributors furthest from the chairman; it is placed at the centre of the table and angled slightly to direct its dead zone between the chairman and the first contributor. A cardioid microphone is used to receive the chairman and his nearest contributor (side A) and is positioned to direct its dead (180°) zone towards the person on the solo side (B). A degree of independent control is then possible, but it is essential to ensure that the cardioid microphone and the (A) face of the figure-of-eight microphone are in phase.

A single cardioid microphone pointing to one corner of a table, as used to cover the chairman and his nearest contributor in the example just outlined, will be suitable for recording the majority of two-person talks. The angle subtended to the microphone will be roughly 90° (45° either side of its centre axis) and the contributors will half-face one another and half-face the microphone; thus there will be a minimum of off-mic quality. This is not an ideal situation but it is a good compromise and can be recommended for use in locations where the acoustics are a little lively.

In lively acoustics the narrower the pick-up angle of a microphone the less will be the reverberation recorded, and a supercardioid microphone may be better suited to lively areas than a cardioid. A point to bear in mind when setting up to record in a lively acoustic is that one should listen carefully before placing the microphone, to establish from which directions there is obvious reverberation.

The microphone should be placed in the least lively area and positioned to present its dead zone to the most lively area of the room. A lively room can be improved by hanging thick blankets or curtains on the walls. Recordists involved in a lot of recording under poor acoustic conditions carry a piece of soft carpet and a few blankets or old heavy curtains as standard equipment.

Where a team of participants is to take part in a televised or audience-attended discussion, it is not possible for the participants to sit round a table in the same intimate manner as is possible for sound-only recordings. It is usual to seat all the

122

participants at a long curving table *with* the chairman or to position the chairman at a separate central table with one group of participants on each side.

Audience-attended discussions usually require some sound reinforcement to enable the audience to hear the discussion clearly, and this presents coloration (reverberation) problems for the sound engineer. However, once a discussion gets under way it is possible to reduce the level of sound reinforcement a little. A gradual reduction in volume is rarely apparent as a growing involvement in the proceedings causes a higher level of concentration to develop, and in consequence the audience is more attentive and quieter.

For quizzes and for larger group discussions it is quite normal to use ten or more separate microphones, one positioned in front (or slightly to one side) of each participant; super-cardioid microphones fitted with anti-breath-blast windshields generally achieve the best overall result.

It is becoming increasingly popular to use tie-clip personal microphones for multi-participant events as they are less conspicuous 'on-camera'. Personal microphones can present their own problems: it is quite normal for participants to pull their beards or to pluck unconsciously at their clothing, and these movements may cause contact noise on the microphones; also, hands and arms may cause acoustic shielding, resulting in loss of quality. When a number of participants are involved in a lively discussion, personal microphones may produce a series of extraneous noises. Super-cardioid microphones, on the other hand, placed half a metre in front of each participant may produce results marred only by the sound of an over-excited (or bored) participant tapping the table-top.

There was a certain television personality who liked to end his show by spinning round three times in his swivel chair before making his exit. To allow this he wore a tie-clip radio microphone, but one week, because of some interference bother, he was obliged to use a wired-up tie-clip microphone and was instructed to 'sit tight' at the end of the show. However, being conditioned to his end-of-show routine he did his usual three spins—and lashed himself to the chair with the microphone lead.

123

MULTI-CHANNEL MIXERS

Multi-microphone set-ups require multi-channel mixers for level and balance control. There are two types of mixer: passive and active. Passive mixers employ resistance attenuation networks on each channel input and all the inputs are connected to a common microphone amplifier. The result is that at full setting every microphone connected to the mixer is electrically in parallel with its neighbours. The resulting input impedance will be low and consequently the hiss level of the amplifier may also be low, but the microphones will load one another—there will be reaction.

When two similar microphones are parallel-connected and receiving different signals, half the energy produced by one is used to drive the other (only when the received signal at both microphones is identical is there no reaction). If two sensitive dynamic microphones are parallel-connected via a length of interconnecting cable, two-way communication is possible: speaking loudly into one microphone produces an output from the other. It is therefore clear that if, say, six microphones are parallel-connected and each is receiving different signals, any one microphone will be driving five others and the output from it will be so loaded as to produce only one-sixth its open-circuit output. To reduce the counteractive effects between microphones, passive mixers should be operated at around the half-full to two-thirds-full mark on the individual input controls, the 'master' level being used for overall control.

Passive mixers are usually transformerless and hence the inputs are unbalanced and not designed to be used with long cables or low-level sounds. In addition, only high-impedance microphones of 5–50kΩ are generally suitable for use with passive mixers. However, they are admirable for tape, gramophone and radio mixing applications.

Active mixers usually have balanced transformer inputs on each channel, and each channel has its own separate amplifier and controls. A comprehensive studio mixer may incorporate a multiplicity of frequency-tailoring, echo, fold-back, pre-hear and group routing controls. Fold-back is a facility for independently routing signals from one or more channels, irrespective of channel level-control settings, to, say, a remote loudspeaker or a

124

pair of headphones. Pre-hear enables an operator to spot-monitor the signals from each input channel individually.

An active electronic mixer does not electrically parallel directly any of the microphones connected to it: as many micro-phones as there are channels available may be used without adverse reaction. With regard to cost, the input transformers alone for a four-channel active mixer can cost more than a complete four-channel passive mixer.

Setting up for recording in stereo

In the set-up for a mono two-person discussion an across-the-table figure-of-eight microphone was recommended. It is poss-ible to use a similar table layout with two figure-of-eight microphones to produce excellent stereo. The microphones are mounted as nearly coincident as possible (one directly above the other); their axes are angled at about 60° and they are worked with their inter-axis centre across the table and positively phased in this direction. The assembly is positioned between the two contributors, not directly on their eye-lines but to one side because, when two figure-of-eight microphones are used for stereo, physical alignment of participants across the assembly results in aural positional collisions in the sound-stage. If the microphone assembly is moved down the table to a position where an angle of 45° is subtended between its centre axis and one of the speakers, the positions of the participants will be fairly hard-left and hard-right in the sound-stage. (Were the inter-axis angle of the assembly set to 90°, the participants' sound-stage positions would be absolutely hard-left and hard-right.)

The participants' sound-stage positions will be governed by their positions relative to the microphone assembly. It will be re-alised that the two across-the-table positions are opposite in phase relationship but identical in sound-stage positions. The reason why one of the participants, in the example just outlined, appears on the left in the sound-stage and the other appears on the right is easy to explain: the sound-stage positions, as stated, are identical on either side of the table; as the two participants face one another across the table, one has the microphone as-sembly on his left, the other has it on his right. Aural positional collisions occur as a result of alignment of two or more partici-

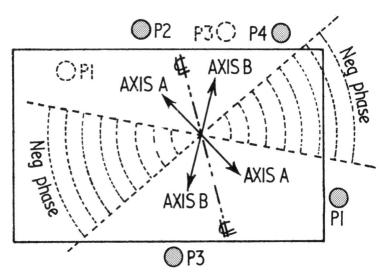

Fig 29 Crossed (coincident) figure-of-eight microphones used to record four
persons at a rectangular table. The resultant sound-stage positions of
P1 and P3 are shown (dotted) to the left respectively of P2 and P4.

pants across the microphone assembly; therefore, if one of the
speakers moved to the other end of the table, remaining on the
same side, the microphone assembly then being in alignment
between the two speakers, their positions in the sound-stage
would be identical—a positional collision.

When three speakers are to be seated at a rectangular table
the same microphone assembly can be used and positioned in
the centre of the table, more or less as before. One participant is
then seated at the centre of one side of the table and the other two
are seated on the opposite side—one well to the left, the other
well to the right (as viewed from the microphone). Thus their
positions in the sound-stage will be centre, left and right respectively.

Two cardioid microphones may be used to cover two-
participant discussions and interviews. The participants take
up positions at one corner of a table; a pair of coincident-
mounted cardioid, super-cardioid or hyper-cardioid microphones with their inter-axis angle at about 60° (a wider angle
may cause the participants to sound too far apart) is directed

across the corner of the table, with the speakers on either side of the corner. The exact position for the microphones will depend on the relative vocal strengths of the two speakers.

In the stereo systems so far described, any movements made by the participants will be realistically conveyed to the sound-stage. If separated microphones are employed, positioned one in front of each participant and electrically positioned in the sound-stage, no movement will be conveyed and there will also be pools of atmosphere at each of the panned positions. These pools of atmosphere are particularly noticeable where the recording location acoustics are rather lively. It is far easier to place a microphone in front of each participant and pan their positions in the sound-stage than it is to set up, and achieve, good genuine stereo.

It is possible to record four people round a rectangular table by using a pair of figure-of-eight microphones mounted in a

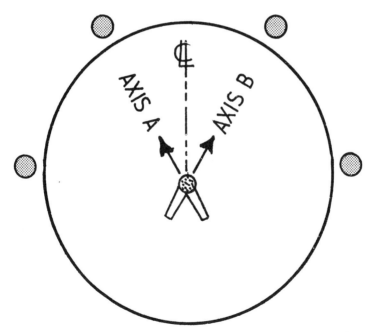

Fig 30 Cardioid microphones set to record four people at a circular table

coincident manner with their axes at 60°, as shown in Fig 29. The participants' sound-stage positions will be in the order P1, P2, P3, P4, left to right. Care should be taken to see that no performer uses the negative phase areas.

A better method for recording four people at a table (or free-standing) is to use a pair of cardioid microphones in coincident mounting (heads in vertical alignment). Fig 30 shows a typical set-up at a round table and Fig 31 shows the equivalent at a rectangular table. The inter-axis angle of the microphone assembly will need to be carefully set to obtain the right positioning of the participants in the sound-stage, and the placing of the assembly should be a little way back from the table centre to achieve good balance. It is better to set the microphone angle roughly (60° is a good angle to start with) and get the sound balance right first, then finally to adjust the angle for perfect sound-stage symmetry.

When even greater angles of sound have to be covered, the three-microphone coincident system, as detailed in the previous chapter, may be used with assurance.

The systems so far discussed have all been operated in conjunction with a table, but equally they could be used without a table. The contributors could be arranged round a stand-microphone system, or they could be seated in easy chairs and

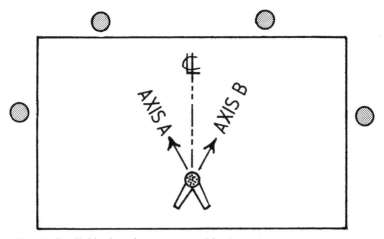

Fig 31 Cardioid microphones arranged for 4 people at a rectangular table

128

use a suspended system or one on an appropriately positioned boom arm.

Some of the details discussed may seem involved and may indicate the necessity of precision setting of microphones to achieve worthwhile results. This is true, of course. Good, genuine stereo does require precision on the part of the recordist. Nevertheless, by understanding what is involved and by following the methods described, even the novice should be able to produce first-class mono and stereo recordings.

7
Recording Drama

Choice of microphone

Just about every microphone polar response ever evolved has, at one time or another, been used for recording drama. In days gone by, when only omnidirectional microphones were available as single units, parabolic reflectors were often used for film sound recording to obtain a sound perspective which was better matched to the picture. It is not always possible to get a microphone close to the action and omnidirectional microphones will pick up camera noise and other sounds associated with picture-making behind the camera.

The situation was eased in this respect with the introduction of the figure-of-eight ribbon microphone which, when held above and in front of the action to be recorded, eliminated most of the extraneous noise by aiming the microphone's dead zone towards the noise—no other microphone has such a good rejection zone. Ribbon microphones, however, are very heavy instruments. Small instruments are now available but they are not very sensitive, for the larger and more powerful the magnet system the more sensitive the microphone.

A hyper-cardioid response allows for a lower working position for a microphone because its rejection zone is at about 130°; it is probably becoming the most commonly used polar response for film work. Many film recordists now use hyper-cardioid and super-cardioid microphones in preference to narrow-beam gun microphones, as they are lighter and easier to manoeuvre (gun microphones being rather long and liable to get into shot) and have better frequency uniformity over wider working angles. A cardioid microphone does not provide its dead zone (180°) in a particularly useful area for working to camera, and in lively acoustics a narrow-beam microphone may give the best results.

WORKING TO CAMERA

This is a very specialised line and requires an understanding of the problems faced by the cameraman. The recordist usually has to compromise between a good sound and an acceptable one. By knowing what the cameraman is aiming for, the recordist can position his microphone for the optimum result without causing difficulties for the cameraman. This may involve continuous repositioning of the microphone to obtain the correct sound perspective to match the picture (as in zooming in or out). The skill comes in knowing the right type of microphone to employ for a particular job and where it can safely be positioned—not only to avoid being in shot but also to make full use of its noise rejection characteristics.

FIGURE-OF-EIGHT MICROPHONE

At one time professional actors were very used to working to a figure-of-eight ribbon microphone, and a technique used by many to give the illusion of retreating from the microphone (fading into the distance) was to step sideways into the dead zone of the microphone while delivering their fade-away lines. This technique produces quite acceptable self-fades in mono but cannot, unfortunately, be used when working in stereo. Now that most drama is recorded in stereo the performers get a little more exercise, because a walk-off actually does have to be carried out to the full.

Drama is still sometimes recorded in mono and it is as well to think about how different polar responses of microphones may be used to advantage. The figure-of-eight response allows performers the freedom to work to either face of the microphone. To work towards one another—to look directly at a co-performer—greatly assists the performers, especially in scenes of an intimate or aggressive nature. A figure-of-eight microphone can comfortably accommodate the work of four performers: two on either face and an arm's length from it. Scripts, where used, should be held in the dead zone (to the side of the microphone) to reduce paper noise. Most actors prefer to stand when delivering their lines and the height of any microphone should be at least chin-height to the shortest performer.

The figure-of-eight microphone is probably the most useful of

131

any for monophonic drama recording, its excellent dead zones being the most important factor. As mentioned before, it is possible to shout in the direction of the plane of the ribbon (edgeways on) and be almost inaudible on the microphone. Therefore, two (or more) people can work at normal conversational levels and at arm's length from the front of the microphone, while someone else quite close by (in the dead zone) is almost shouting. This may be required when trying to convey the illusion of, for example, being with (and listening to) two persons at a conference, while another individual is holding forth from the platform, in a loud voice.

EFFECTS

Special effects—like pouring a cup of tea and adding milk and sugar, or winding up a clock—can be carried out at the microphone used by the artistes and simultaneously with their speeches. Any effects required to sound distant can be performed somewhere between the front and edge, depending upon the perspective one is aiming for. Many recordists use a separate microphone for sound effects as this gives the 'special-effects man' greater freedom to do whatever he has to do. This presents its own problems, however, because another microphone on the set automatically picks up some of the speech material concentrated around the performers' microphone and there is a resultant increase in reverberation; also, if the two microphones are some distance apart, distinct echoes may be heard.

Where effects of a dangerous nature are required in a production, eg the breaking of windows and bottles or the crashing of falling objects, great care must be exercised. These effects should not be performed 'in the run' but recorded separately and played-in during the performance. The set should be cleared of onlookers and a very large dust sheet laid to catch splinters of glass and so on. After the diabolical or magnificent effect is achieved and the dangerous remnants disposed of the set should be little worse for the experience. Everyone involved in further use of the set should be warned of the danger of glass splinters possibly remaining about the set.

OMNIDIRECTIONAL MICROPHONE

Omnidirectional microphones will not generally perform too well for drama unless the reverberation of a set (or room) is to be used for dramatic effect. Because of its all-round pick-up, it will produce exaggerated perspective when used in reverberant locations and this characteristic (better avoided for talks and interviews) may be just right for producing dramatic sounds in large rooms, corridors and caves.

CARDIOID OR SUPER-CARDIOID MICROPHONE

Cardioid and super-cardioid microphones are readily available and generally cost less than ribbon figure-of-eight micro-phones; they are also quite suitable for drama use. Incidentally, anyone who has two identical cardioid microphones has effecti-vely a figure-of-eight response at their disposal because two back-to-back mounted cardioids (one head directly above the other but facing the opposite direction) connected in anti-phase (direct or via a mixer) will produce a good figure-of-eight response. In fact, I would recommend enthusiasts to obtain two good cardioid or super-cardioid microphones as their first instruments, for they have the most versatile of any polar response.

With two well-matched cardioid or super-cardioid micro-phones an enthusiast can produce first-class stereo. He can also arrange them to provide a single figure-of-eight response and use the assembly in the horizontal mode for its double working areas and effective dead zones, or in the vertical mode for cancel-lation of horizontal ambient noise when interviewing or record-ing special effects. When used singly, cardioids and super-cardioids provide good isolation from sounds at the rear of the microphone (the front-to-back ratio of a good cardioid is 20dB). A cardioid microphone used vertically provides an all-round working angle: it can be used facing vertically down-wards or upwards with performers spaced all round it.

Location

It is not always necessary to use the exact physical location in order to convey the quality of sound associated with a particular site; one of the reasons for using, whenever possible, a location which exhibits the desired acoustic is that in natural, large acoustics performers at different distances from the microphones sound quite different in perspective and degree of reverberation.

As a distant door opens in a large acoustic and a person enters and approaches the microphone, there is a definite change in sound quality. At the door, the voice of the performer is very coloured in acoustic reverberation; then, as the person progresses towards the microphone, the sound quality of his voice varies with every step forward because each new position excites slightly different frequencies of reverberation and there is also a continuously varying ratio of direct to reflected sound. When the performer is close to the microphone the coloration will be low. This is precisely the effect noticed in an actual room and is, therefore, the natural sound. I have not yet heard to my satisfaction any electronic device which is capable of simulating the acoustics of a large room, but with the advances being made in microprocessors it may soon be possible—time alone will tell!

Being lucky enough to work in a radio drama studio with many different natural acoustics, I hardly ever have to resort to electronic echo devices for acoustic enhancement. I do use them for creating special effects, though: creaking door hinges, dogs howling after midnight and mysterious footsteps usually become a little more dramatic with some electronic embellishment!

Stereo drama techniques

Stereo microphone techniques are very little different, as far as the performer is concerned, from those used in mono, except that there is no dead zone to produce self-fades. In stereo drama it is far better that the performer actually carries out the moves or actions to be conveyed to the listener, as direction of movement and perspective are essential. Perspective will be exaggerated when the performer moves from a near microphone position to a more distant position, because it is usual for performers to work at an arm's length from microphones. Normally when two persons are conversing they stand at almost double arm's length from one another (unless environmental noise is high). Therefore, when moving back from a near microphone position to a more distant position, the apparent distance conveyed is about double the actual distance moved. When the near position becomes about a metre from the microphone the actual and the conveyed distances of more distant positions become more comparable. Were we all to hold our normal day-to-day conversations at arm's length we would come to accept a larger perspective ratio as normal.

The illusion of distance can be conveyed by utilising the acoustic reverberation of a location where, as already mentioned in the mono section, the dead zone of a microphone is used to distance a performer by variations of direct to reflected sounds. It is difficult to use the dead zones of microphones paired for stereo. In the case of figure-of-eight pairs the dead zone of one microphone coincides with the axis of the other and vice versa. A pair of coincident super-cardioid microphones with their inter-axis angle at 60° provides about the best stereo working dead zone of any system and is also capable of producing excellent stereo.

THE FADE-BOARD

A distancing method used in old-style 'live' stage shows can also be employed in modern drama recording. The technique, which was and probably still is used in some theatres to fade the sounds of special effects and to accomplish simultaneous (or separate) fade-outs of an individual or group of performers and fade-ins of another, is achieved with a very simple mechanical device known as a fade-board. If, while addressing an audience or a microphone, a performer has a piece of board moved slowly in front of his face, there will be a great reduction in his delivered volume.

A fade-board should be about 1m square and one edge should have eight or more deeply-cut V sections, so that the filtering effect is gradual as the board is introduced between performer and microphone. To make the fade even more effective, the side of the board facing the performer should be covered in some sound-absorbing material, such as thick carpet felt. If you have not already tried a fade-board I do strongly recommend it: you may be surprised by its effectiveness.

OTHER EFFECTS

In a production where a number of people were required to sound as though they were in a garden and shouting down into a well, a 3m length of large-diameter plastic soil-pipe was found to be quite useful. Two small cardioid microphones were positioned in the tube, one at the blocked end and one about halfway down. Their outputs were taken to the mixer and positioned stereo-left and stereo-right; they were also fed to a tape recorder which, together with some electronic echo, produced a delayed, flutter echo. A second pair of microphones was positioned above the open end of the tube—the tube being supported at this end by a chair—so that the performers were able to deliver their lines to the 'garden' microphones and (by bending down slightly and working close to the open end of the tube) to the 'well' microphones. With the addition of a few echoey dripping water sounds the total effect was most dramatic.

In another production, stage thunder was required and this was produced by rolling coconuts down a long sloping sheet of thin metal supported by a few pieces of old floor-boarding. Corrugated-iron roofing pieces laid sideways would probably produce even better results than the flat sheet which I used.

Special effects which cannot be produced on the set will need to be recorded at some other locality and played into the scene directly or replayed to a loudspeaker on the set. It is generally better to replay recorded effects through a high-quality loudspeaker on the set because then the correct position, perspective and loudness can be accurately set to produce the same effect that the live sounds would, were they to emanate from that position.

Grandfather clocks will sound most natural when delivered via a high-quality loudspeaker on the set, and I do not think anyone could tell whether the sound came from a loudspeaker or from the actual clock. Incidentally, the volume level of clocks and similar scene-setting pieces is usually reduced once the effect is established and the scene gets under way, because clock ticks tend to become rather obtrusive after a while; they can always be reintroduced at a higher level when the occasion demands it—in moments of suspense.

136

Many a laugh can be had when using a background clock. When rehearsing a scene of an intimate or tense nature, not many performers are able to carry on with a straight face when the gentle ticking changes to a sudden clang, bizzzongkyty-wizz of springs and then reverts to a sedate tick-tock. No one appreciates things overdone—subtlety wins the day. If just one tick and one tock are replaced with alternative sounds, a few seconds may elapse before the effects register with the performers—but register they will. (I hope I have not sown any prankish ideas into the minds of anyone who is about to make up the sound-effects tape for a local amateur dramatics society play.)

When required for replay via loudspeakers on the set, all recorded sound effects should be as free from ambient sounds or recording-location reverberation as possible otherwise, whenever the sounds are replayed, the ambience of the recording location is also heard, which may not be acceptable.

FIGURE-OF-EIGHT ASSEMBLIES

Returning once more to microphones, the stacked figure-of-eight microphone assembly is very suitable for drama recording. It provides two positionally identical working zones, and the same techniques as used for talks and discussions can be used for drama. Remember, though, that aural collisions occur when there is a direct alignment of microphones between two performers, and that the two out-of-phase side zones must not be used. To assist the performers the floor may be marked to indicate the areas to be worked and the areas to be avoided. I do not find it necessary to mark the floor except, occasionally, the prohibited areas when a large number of artistes are performing together in a scene. I find that most of the artistes with whom I have the pleasure of working have worked with stereo before and understand the complexities, and those who are new to stereo usually learn very quickly and write 'move downstage left', etc, on their scripts.

Figure-of-eight assemblies should be angled to suit the scene. When a large number of performers work fairly close to the microphones the inter-axis angle should be reduced from the 90° maximum to somewhere nearer 40°; this will provide two much wider working zones (140°) and appropriately reduced out-of-phase zones (40° each). Figure-of-eight assemblies are very useful for effects too: while artistes are performing in one of the working zones, effects can be dealt with in the other zone without the operator getting in the artistes' way, or vice versa. When an effect is required to coincide with the position of one of the artistes on the other side of the assembly, it is necessary only to cause it to occur on the aural collision path—on the line of artiste/microphone alignment. To prevent unnecessary reverberation being picked up or to prevent the sound effects from radiating about the location, the whole of the effects area can be screened in.

Acoustic screens are used extensively in sound drama. They can be used to build rooms, to modify the acoustics of existing rooms and to provide an isolation barrier between one area and another. Acoustic screens are usually about 2m high and just over 1m wide (some have add-on units to bring their height to about 3m) and they stand on wheels or plain bracket stands. The most common type has one side padded with sound-absorbent material and the other side plain and reflective. Other types can be obtained: some have both sides padded; some have both sides plain; some are fitted with 4mm sheets of clear plastic; some have hinged sections to convert the clear reflective panels to sound-absorbing panels. Roof sections are also obtainable. When a small, boomy acoustic (similar to a telephone-box) is required, four screens are positioned in a square with reflective sides inwards. A small gap is left between two of the screens to allow access. If clear plastic screens are used the result is quite similar to a telephone-box in appearance and in acoustic. In a small booth a pair of super-cardioid mics will produce good results.

In scenes requiring a telephone voice to be recorded, a standard telephone can be used. A microphone can be held against the ear-piece of the handset, or a magnetic pick-up (held to the telephone by a suction cup) can be used. Both methods receive a rather distorted signal but nowadays, with the almost daily broadcasts of telephone-quality interviews, most listeners can hear what the voice at the far end is saying. In high-quality stereo drama, however, the contrast is often too great and it is necessary to compromise between the actual telephone quality and something which can be clearly heard—even by a hard-of-hearing listener using an off-tune transistor portable with almost flat batteries!

It is easy to limit the frequency bandwidth to the passband of the telephone system (300Hz–3kHz) but that does not satisfy the hi-fi stereo listener, and on a small clapped-out portable it probably sounds no different. What is needed is a degree of frequency coloration, quite a lot of noise and a modicum of distortion. This can be achieved by using a small loudspeaker unit (of a quality you would be ashamed to admit to your friends that you had actually bought) instead of a frequency-limited microphone or an actual telephone. A moving-coil loudspeaker of about 7cm diameter and of uncertain lineage will require an enormous signal amplification, which will result in a high degree of noise and distortion—just the effect needed in fact: not as distorted as a real telephone and not as clear as a frequency-limited high-quality microphone. A passband of 300Hz to about 5kHz will probably be best for producing the desired result.

When controlling levels of frequency-limited sources such as telephones, the sound level should be balanced by ear, not by meter. If narrow-passband telephone-quality speech is allowed to register to the same level as normal high-quality speech it will sound more than twice as loud as the high-quality speech, though it will depend a lot upon the type of level meter used. To modify the frequency response of a microphone (or small loudspeaker or headphone ear-piece used in reverse) try a small length of cardboard tube placed over it—or a funnel-shaped piece of plastic or a tin can.

When microphones, particularly those with figure-of-eight and hyper-cardioid polar responses, are placed close to walls or screens with reflective surfaces, the reflections will modify the quality of sound. In the examples of the telephone-box and the cardboard tube these reflections were used to advantage but in normal use hard reflections from surfaces too close to the micro-

139

phone should be avoided. If reflections are found to be particularly troublesome it is possible for you to modify the reflective surface by adding something highly absorbent, such as thick carpet felt.

HYPER-CARDIOID MICROPHONES

Hyper-cardioid microphones are also excellent for drama work, although the rear area will not be as useful as in the figure-of-eight system: the positional guide of performer/microphone alignment, so easy with the figure-of-eight assembly, is not available to the same accuracy with the hyper-cardioid. There are two out-of-phase zones to avoid when working with hyper-cardioids, and in a pair mounted at 65° these zones are between approximately 90° and approximately 150° on either side of the centre working axis.

SUPER-CARDIOID MICROPHONES

When mounted as a coincident pair at about 75° inter-axis angle, these are excellent for stereo drama recording. Performance work must be carried out in the front area only, but with a working angle of roughly 200° no problems should arise in positioning a number of artistes round the microphone assembly. There is no absolute dead zone but there is a zone of about 60° which is fairly insensitive. The directional (sound-stage) positioning of sounds originating in this relatively dead zone will, however, be somewhat vague and uncertain and the zone is not to be recommended for sound-positional use. If the zone is used for some dramatic effect in stereo and at some time the result is possibly going to be heard in mono, the mono/stereo compatibility should be checked. Used in the centre of a highly reverberant location and at a given working distance, a super-cardioid system will probably produce the clearest and least coloured result of any microphone system—with the possible exception of a pair of gun microphones.

Cardioid microphones are commonly used in stereo drama but I prefer super-cardioids: I think they produce more precisely defined positional results than cardioids and they have a better control over perspectives and reverberation. Cardioids, however, afford the widest working angle of any system and

140

there are no phase problem areas. A working angle of up to 300° is possible with a pair of cardioid microphones when their inter-axis angle is reduced to about 45°.

A problem which often arises in sound drama when the actors are required to work close to the microphone is that movement about the centre line becomes very much exaggerated. When several actors have to work close to the microphones their positioning can appear to the listener to be cramped on either side of the sound-stage, while at the same time producing exaggerated movement about the centre line. This problem is easily overcome by using the three-microphone stereo system. Referring to Fig 28, on page 112, microphones A and B are the wide-angled stereo pair of the system: on their own they produce stereo of a kind which, for the 'hot seat', suffers poor centre definition and bunching of sounds on the extremes of the sound-stage. When the left and right microphones, A and B, are faded out and the centre microphone is brought into operation, there will be centre mono only; then as the left microphone is reintroduced there will be stereo from sound-stage left to centre, and likewise on the right when B is reintroduced. Microphone C is the centre filler and image stabiliser and microphones A and B are the outriders. As the level of C is increased in relation to that of A and B the image width is reduced.

Twelve or more actors can work around a three-microphone system. They can use an angle of about 230° and, if the result is to appear to be a tightly packed group, all that is necessary is to balance C against A and B (generally a three-finger operation). When the actors in a two-person scene work too far apart, worry not—no need to rush to the set to reposition them—just waggle the middle finger and, 'hey presto', they are repositioned electronically.

To provide very firm side images to the sound-stage, and thus improve the stereo effect for the side-seat listener, it is necessary to use fully the very wide working angle of the system. If action is arranged to take place in a position some 115° from the centre line, it will happen in the dead zone of one of the outriders and will give maximum separation between left and right; therefore the image will appear very firm on the appropriate side of the sound-stage. The system offers its best stereo-image performance (quite apart from the image-width control facility) when all three microphones are working at roughly the same gain settings.

Another way in which a great number of actors can be made to sound very tightly packed is to space them round the large angle as before, but this time, instead of using the centre microphone to decrease the image width, use it to double the grouping. Pan microphone A to the left; pan microphone C not to the centre but to the right; then pan microphone B to the left, so that there is a fold-over of the field sound. As will be readily appreciated, a three-microphone system offers you the opportunity of very flexible and complete control over the wide working angle available.

GUN MICROPHONES

These are not necessary for studio drama, although they can be used for exterior and stage work. When a pair of super-cardioids is positioned on a stand just in front of a stage to record a stage play or similar performance, there will be very large variations in recorded perspectives between downstage and upstage positions. These variations are not apparent to anyone in the audience in the main body of the auditorium, because no one is normally able (or would wish) to sit that close to the stage area. Ideally, the microphone should be brought back to a point where the proscenium arch subtends an angle of about 90° to the microphones—but this suggests placing them over the middle of row six in the auditorium. A pair of gun microphones placed that far back and mounted (at 30°) just below the centre lighting barrels would produce surprisingly good results. There would be phase errors which would produce a slightly variable sound quality in mono but, compared with the phase problems resulting from the use of four or more spaced microphones by the footlights, these would be minimal. The gun microphones would receive the right perspective ratio of upstage and downstage sounds and also a degree of theatre ambience. I would recommend to anyone involved in theatre stage recording, and who has access to them, to try a pair of gun microphones as suggested—I am sure that they would find the results to be very pleasing.

With very large stages it will probably be necessary to use two super-cardioid pairs spaced some distance apart along the stage front, supplemented by a pair of gun microphones. The left microphone of the stage-left super-cardioid pair should be placed stereo-left in the sound-stage and the right microphone of the stage-right pair should be placed stereo-right. The right microphone of the stage-left pair and the left microphone of the stage-right pair should both be panned to sound-stage centre. These two pairs will cover the wide downstage area without a great deal of phase interaction. To cover the middle-stage and upstage areas the super-cardioid pairs are supplemented with a pair of gun microphones fixed high up above the centre of the proscenium arch on the stage side. By carefully balancing the gun microphones with the super-cardioid pairs, this depending upon the area of the stage from which the sounds originate, both good perspective and good positional detail can be conveyed to the sound-stage.

Gun microphones provide a useful stereo working angle of 70° and, when used for recording drama, artistes need work no closer than about 3½m from the microphones to produce close-perspective sound. Therefore, gun microphones are excellent (allowance being made for phase errors) for covering events where much movement is involved. I have used a pair of gun microphones to cover open-air drama, song and dance and, except when the performers were projecting away from the microphones, very good results were achieved. Perspectives will not be exaggerated when gun microphones are used for exterior drama work. In fact, because the pick-up angle of the system is equivalent to the angle of the listening sound-stage, normal full-length approaches and departures can be recorded with great realism. For external use 28° is usually the inter-axis angle which produces optimum results. Where special effects are not available at the time artistes are recorded they can be added later in a copy/mixing operation.

A point to bear in mind when collecting special effects is that the sounds produced by an actual object do not necessarily convey to listeners the effect intended. Other objects manipulated in various ways may be found to produce an even more realistic sound than the genuine article.

FADING

Drama scenes which have atmosphere added to convey that the scene is exterior (or general hubbub to convey an interior scene—a public house for instance) can be end-scene faded in two distinct ways. One method is to fade dialogue and effects together, using a master fader on a mixer. The other method controls each separately: the dialogue is faded normally but the fade of the effects or atmosphere is delayed slightly. At a point where the dialogue is at about half-level, the atmosphere is progressively faded with the dialogue but at a slower rate, until finally they both fade out together.

The first method, the simple way, usually produces a fade where the atmosphere leaves the scene ahead of the dialogue. The reason for this is that the atmosphere in a scene is usually very much lower in level than the dialogue—in fact it is already at a level not far above the minimum which a listener is able to hear in the environment of a quiet listening-room. Therefore, only a few decibels of level reduction will cause the background effects to be masked by the fading dialogue level, the ambience of the listening-room and the noise of the reproduction system—and they will be inaudible. The artistic approach to fading is far better in that it produces a scene fade where both the dialogue and the supporting effects are heard to fade together.

LOW-FREQUENCY ATTENUATION

Male voices, even the most resonant, do not have any useful frequencies below 100Hz and when deep, resonant voices are working at normal arm's length from a microphone the result is usually an accentuated, boomy quality. Drama recording benefits enormously, in my experience, when frequencies below 100Hz are severely attenuated. I always use a steep-cut bass filter (80Hz 18dB/octave) on all microphone circuits. Practically all floor and airborne rumbles are cut and the whole ambience of scenes is generally improved. When listeners are not subjected to continuous low-frequency rumbles from air-conditioning, traffic, etc, the low-frequency content of special

144

effects like thunder, guns, bombs and big bass drums has a much more dramatic impact. The desirability of cutting out low-frequency noises does not only apply to drama but equally to talks and discussions, wildlife and music recording.

8
Recording Music

Music recording can be divided into three classes: (i) simple, single-microphone recording; (ii) multi-microphone simultaneous recording; and (iii) multi-track recording, where the final result may originate from a build-up of several separate non-simultaneous 'takes'.

Single-microphone recording

Taking the simplest first—the single microphone (for our purposes this includes two or more coincident microphones for stereo)—we shall examine some of the conditions for achieving good results on a variety of materials.

SOLO SINGER

Recording a solo singer may present the same problems as outlined previously when I referred to the use and abuse of microphones for sound reinforcement. I mentioned that undisciplined performers often use a microphone as a dancing partner; obviously the recordist will not be able to achieve satisfactory recordings from a microphone so mishandled. A separate microphone positioned about 2m in front of the performer will probably be more satisfactory. If the soloist does require something to wrestle with, a separate microphone with a lead which goes nowhere will probably help enormously! Recording direct from a microphone hand-held by an experienced recording artiste is quite a different matter. An experienced performer balances the voice level automatically by holding the microphone at a greater distance when hitting the louder notes and reducing the distance on the weaker ones, and at the same time avoids breath blasts by holding the microphone to the side of the mouth.

SINGER WITH ACCOMPANIMENT

When a singer is to be accompanied by several musical instruments or by other singers, a good recording can be achieved using only one microphone position if the performers are correctly balanced.

I find the quickest and easiest way to balance the sounds from a number of performers is actually to stand at the point where the microphone is to be and while the artistes are performing indicate in some way to each in turn whether to advance or retreat. When I am satisfied that a good aural balance has been achieved, I dash to the control point to verify the result. This method is very quick and it causes no unnecessary concern to the artistes, for they are not interrupted and moved several times as happens if the balance placing is directed solely from the control point. Balancing for stereo sound-stage positions can also be aurally judged from the floor, because familiarity with the performance characteristics of the microphone chosen for the recording enables one to assess correctly the position of a performer on the floor (or stage) for a given position in the sound-stage.

SINGER WITH GUITAR

Here the position of the microphone will depend on the relative volume levels of the singer and the guitar, also on the type of microphone employed and its elevation and angle. For instance, if the relative levels of singer and guitar were equal and a cardioid microphone were to be positioned equidistant between the two, the pick-up of voice and guitar would be equal. If, however, the microphone were angled upwards and on-axis to the singer in such a manner that the angle of the microphone to guitar was 90°, the pick-up balance of singer/guitar would be 2 : 1, because a cardioid is −6dB at 90°. If a hyper-cardioid replaced the cardioid microphone the balance for singer/guitar would be nearly 4 : 1. It may seem unbelievable that a small change in microphone angle can have such an effect on balance, but it is so.

ORGAN MUSIC

Organ recitals performed in churches are usually simple to record, but the main problem is one of tonal balance. Larger

147

organs may have some of their pipes in far removed quarters, often in very high and inaccessible places. Positioning a microphone high enough to receive a reasonable amount of direct sound from the high-frequency pipes is a common problem. Normal stands are nowhere near tall enough for organ recording. A height of 5m is by no means too much; in fact, it may be necessary to go to 15m or higher in cathedrals. It is often possible to sling-mount microphones quite high up in cathedrals by suspending them from the galleries. It would be wise to listen carefully to the organ before deciding on rigging positions. If it is possible to listen from a high point the advantage of a fairly high microphone position may be obvious. On the other hand, some vibrations of windows or panelling, not audible from below, might be clearly audible higher up.

The type of microphone to use may depend upon the available mounting positions. Likewise, the choice of position may depend on the type of microphone available. For churches of a relatively 'dry' (not very reverberant) acoustic a figure-of-eight microphone (or a coincident pair of these for stereo) will probably produce the best result, once mounted in a favourable position. In recording an organ for specific purposes, eg to use in scenes in a recorded drama, the position chosen for recording will depend on the perspective demanded by the scene. If it is to give the illusion of taking place in the vestry, the organ is best recorded from the vestry to get the correct perspective.

CHOIRS

Recording choirs in churches is relatively simple, but just to place a cassette recorder on a pew and a microphone on a *Book of Common Prayer* is unlikely to pick up the diction of the choir well enough! The choir's best balance and the clearest diction can usually be heard at a point about 1m behind and 1m above their conductor, necessitating a rather tall microphone stand. The conductor usually stands on a dais, and the height of this must be taken into account to ensure that he does not acoustically shadow the microphone.

An alternative to a tall stand—and one far better for recording audience-attended performances—is a neatly slung microphone. Microphones can be slung in advance of the arrival of a

148

visiting choir, but the choir may propose to sing from an area not served by the microphones. This problem then presents the recordist with the choice of persuading the choir to perform in the pre-rigged area, or of getting to work on repositioning the microphones. In churches, and in cathedrals in particular, the pillars are too large in circumference to allow last-minute re-rigs of slung microphones, unless a convenient quick-fix method can be devised. Ropes around pillars are not easy to fix because ladders are usually required to position the ropes high enough up the pillars.

A device for quickly attaching to, and slinging microphones from, the pillars in churches is shown in Fig 32. It is adjustable, to attach safely to the small, vertical round-section shafts which run up the main columns and arches in most churches built in the traditional carved-stone style. By the use of these clamps the microphones are suspended across the nave, choir or chancel. If the wing-nut on the clamp is not too tight, the unit can easily and safely be pushed up the pillar to the required height with a long broom handle. Attached to the bottom of each unit is a length of cord, and it is necessary only to pull the cord and at the same time apply a little side pressure on the unit (with the broom handle) for the clamp to slide easily down the shaft without marking the stonework. The edges of the unit must be chamfered to prevent catching the joints in the stonework.

With a pair of these simple devices a relocation, or change of height or angle of microphones can be accomplished in minutes.

Figure-of-eight, hyper-cardioid, super-cardioid and cardioid microphones are all suitable for recording choirs. The choice will depend on the acoustic properties of the location. In a very lively location it may be necessary to use super-cardioid microphones to reduce the amount of reverberation pick-up, whereas in a 'dry' acoustic a figure-of-eight microphone spaced further from the choir may produce the best result. In this respect the recordist using a modern microphone with a variable polar response is, indeed, very fortunate.

Single-microphone recordings of, for instance, brass bands should be dealt with in the same manner as choirs. The balance of choirs and of brass bands is the responsibility of the performers and their conductors—not the recordist—though he

149

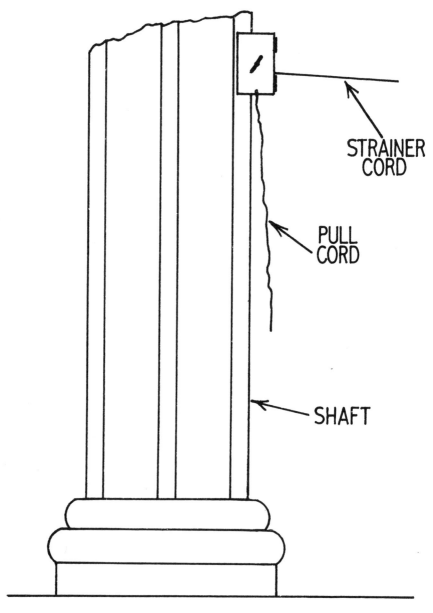

STRAINER
CORD

PULL
CORD

SHAFT

Fig 32 A simple, quick-fix device for attaching to shafts

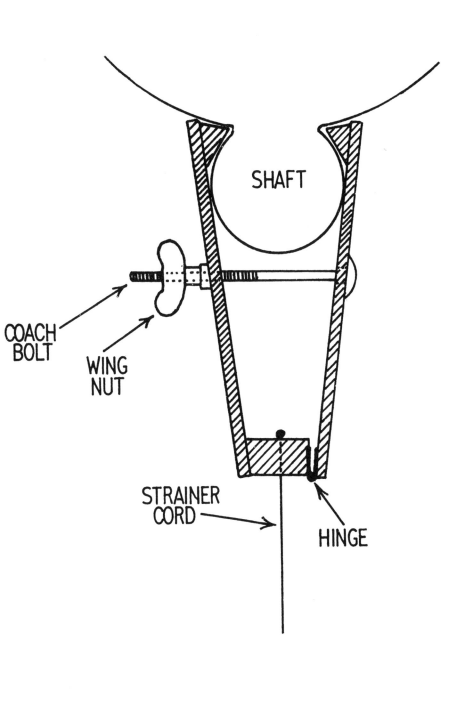

SHAFT

COACH
BOLT

WING
NUT

STRAINER
CORD

HINGE

may be able to offer suggestions as to how a better balance might be achieved, such as moving one or more over-enthusiastic singer to the back row.

PIANO MUSIC

The position to place a microphone for recording piano recitals will depend on the acoustic liveliness of the location, which will also influence the choice of microphone. The lid of the grand piano will probably be raised to its maximum opening and this, being at about 45°, will reflect the sounds in a mainly horizontal direction—towards the audience. To achieve a good tonal balance, the best position for the microphone will probably be some 2–3m away from the piano, on the open (audience) side, looking downwards at an angle of 25°.

Stage and platform floors are subject to intense low-frequency vibrations and some form of mechanical resilience should be placed under the feet of non-sprung microphone stands to reduce bumps and rumbles. Slung microphones are probably the best arrangement because slings provide complete isolation from floor vibrations. Be careful, though, what you tie the sling to. I once tied a microphone sling to a water-pipe in a public hall. All very convenient, but unfortunately, when the central heating was turned on later in the day the pipe shook with pulsations from the water-circulating pump, and the noise was transmitted along the sling to the microphone. Fortunately, this was discovered before the audience arrived and while there was still time to do something about it. The cure was simple, in fact: a wooden door-wedge was inserted between the wall and the pipe at the point where the sling was tied, and the vibrations then ceased.

Two of the most continuous and, to my ears, distractingly unpleasant noises are ventilation rumbles and heavy traffic rumbles. These are particularly noticeable and often intrusive during quiet piano passages. In compositions which do not call for the use of the lowest octave, surely a recording benefits enormously from the introduction of a 60Hz steep-cut filter? Music-recording purists may view any form of frequency restriction as taboo, but I am not a purist—I am practical and my aim is to produce the best results in whatever form of recording work comes my way. If the quality of a recording is improved by using a filter, by all means use one.

Orchestral concerts are probably the easiest of any musical performances to record. There is a traditional and internationally accepted layout for orchestras and rarely does one find any radical deviation. The reason for the rigid adherence to this traditional layout stems from centuries of fine judgement in orchestral balancing. If a violin were able to produce the same ear-splitting volume as a trombone, no doubt violins would have been placed at the back. Every conductor endeavours to maintain the very fine balance which this traditional layout affords.

The only control the recordist has over the final balance is to position his microphone correctly. The microphone should be positioned behind the conductor where an angle of about 90° to the extreme side players is subtended to the microphone (unless the acoustics of the location are very lively, in which case the microphone will need to be much closer). A satisfactory height will be where the angle of the microphone, as viewed from the front line of the players, is about 35° from the horizontal. An omnidirectional microphone may produce good mono results in a rather 'dry' hall, and 90° coincident figure-of-eight microphones may produce excellent stereo results.

Watch out for the ventilation outlets when positioning microphones—they may not be in operation at the time you are rigging but there could be quite a breeze when they are on during a performance.

For recording large-scale open-air musical performances, my first choice of microphone would be a gun microphone and my second choice a super-cardioid. The gun microphone would need to be on a long pole, as it could be impossible to get through the assembled crowd; anyway, to get to the front would only result in a poor balance— better to get the microphone a few metres away but high up and angled downwards. For smaller musical events of, say, six performers I would use a pair of hyper-cardioid microphones for stereo and a super-cardioid for mono; also, I would have them in a special windshield on a long pole.

153

Multi-microphone recording

Additional microphones may be required to supplement the main ones and to reinforce particular sections in, for example, an orchestra. Alternatively, separate microphones may be used for each and every sound source to be recorded. In the latter, the signal level of each microphone is controlled separately via a mixer to produce a balanced sound. Electronic echo is added to enhance the overall sound.

SINGERS

Where several singers are to entertain an audience and their performance is to be recorded, it may be more convenient for each singer to work to a separate microphone. If the recording is in stereo, the recordist may decide to produce a more distant perspective on the singers. (Working so close to their individual microphones the singers' perspectives will be very close.) For this, a pair of super-cardioid microphones can be positioned just forward and above head height of the singers (how far forward will depend on how many singers there are) to where a 90° angle is subtended between the two side singers of the group and the microphones. This will be the main system for sound-stage positioning, for improved perspectives and for more ambience. The individual microphones will be panned electrically to sound-stage positions corresponding to the positions of the singers reproduced by the main pair. The close microphones may be used to reinforce the main microphones where this becomes necessary (if the balance has changed since it was organised for the main microphones), or they may be used to 'pull' a lead singer perspectively more forward.

STEREO AMBIENCE

For stereo recording, any ambience (audience) microphones should be fixed in a position which achieves good ambience balance in stereo. Individual, isolated, panned ambience microphones produce mono blocks in the sound-stage.

Recitals recorded in locations which are relatively dry in acoustic reverberation may benefit from the addition of ambience microphones, these positioned some distance from the primary microphones and from the recital instrument. In a church where the organ is at the east end, a pair of cardioid ambience microphones placed half-way down the nave and directed upwards to the west may help considerably to produce a more 'churchy' sound.

Figure-of-eight microphones probably produce the best stereo ambience of any system but the mono signal (the combination of the two stereo signals) may still sound too dry. With figure-of-eight stereo microphones nearly half the ambience pick-up is negative phase. This negative-phase ambience, together with the positive-phase ambience, may produce a rather grand wide-stereo ambience, but phase cancellations will reduce the ambience level to meagre proportions in mono and you will find it is often necessary for you to overdo the stereo reverberation slightly in order to achieve the right effect in mono.

CHOIRS IN STEREO

When recording choirs in stereo, the sound-stage side-slip experienced by the off-centre listener can be minimised by reinforcing the sides of the sound-stage. The three-microphone system will be of great assistance in producing better stereo if the choir can be arranged to present a semicircle to the system. When this can be accomplished, although an improvement may not be apparent from the 'hot seat', the sides of the sound-stage will be held firm and a very much better stereo effect will be presented to the off-centre listener.

Where the conventional straight-line choir arrangement has to be adhered to, additional microphones may be positioned fairly high, just forward of the extreme sides of the choir and directed down towards the back row. These side reinforcers deliver their signals to the appropriate sides of the sound-stage and help to stabilise the side images. However, if these microphones are used at too high a gain level the 'hot-seat' listener will receive impaired stereo because blocks of mono sound at the sides of the sound-stage will result.

The same techniques can be applied to record orchestras. Separate, single microphones may be positioned over the violins on the left and by the basses on the right. Individual microphones may also be used to bring up the strength of other sec-

155

tions in an orchestra and to pick out a solo passage here and there.

The polar response of every microphone used for reinforcement must be carefully considered, and the positioning of these microphones is most critical. It is not too difficult to discover in which direction a particular instrument emits its sound, and when a microphone is being set up to reinforce a player or section it should be positioned in such a way as to minimise the pick-up from other players or sections. Also, for stereo recording, all 'spot' microphones must be carefully panned to coincide with their main-microphone sound-stage positions, otherwise positional shifts will occur when the spot microphones are faded up.

Another multi-microphone recording method dispenses with the main microphones, employing individual microphones for each and every section in an orchestra, or for each player in the case of small bands and rock groups. A drummer may have up to ten microphones positioned round his drum kit. Each microphone will have to be carefully chosen, positioned and angled, and fed to a mixing console. The operator-recordist or balance engineer will adjust levels, stereo positions and amount of echo to achieve a result which is well balanced and pleasing to listen to. The sound-stage positions will probably be totally different from the actual positions of the players in the studio, especially if the studio is rather cramped. The recordist may find it necessary to place tall screens between some players to reduce 'spill' of sound from section to section. Percussion may spill over the entire studio and have to be completely screened in.

GROUPS

Drummers of pop groups certainly need to be isolated or screened off from others in the group because the sound level they develop is often above the threshold of pain (to my ears anyway); this very high volume not only spills all around the studio but can often be heard out in the street. Anyway, dynamic (moving-coil) microphones cannot overload at very high levels as can many a capacitor. An electret microphone with a maximum sound pressure level of 105dB can be expected to produce more distortion than signal when placed in a bass drum.

One can easily run short of enough microphones for a full drum kit, because once one starts to use a separate microphone close to one particular instrument it is surprising how it shows up other inadequately covered instruments. However, with a drum kit a compromise has to be struck: where possible, one microphone has to be positioned (without being in the drummer's way) between two or possibly three items of the kit, thereby reducing the number used. Other members of the group may be provided with individual microphones or they may be able to share.

Sound balance of groups is very much a matter of personal choice and current fashion. There have been many balancing fashions down the years: at one time it was fashionable to have only the minimum of bass and drums, the rhythm being pro-vided by piano, trombones, saxophones or even mouth-organs, and lead singers were heard above a nicely balanced rhythmic backing. One of the present fashions in balance seems to be to drown the singers in a cacophony of over-heavy bass guitar and drums. The fashion will inevitably change again one day: the drummer may move to a more distant perspective, while electric guitars may give way to saxophones and accordions—who knows!

Multi-track recording

Multi-track recording is used to compile a unified whole from a number of separate sources. The balance required for the final mix may not be known; therefore, every sound is recorded at maximum levels on separate tracks. Multi-track recording machines with 8, 16, and 24 tracks are available. The material is built up by first recording, say, a rhythm section of three players. The tape, which for a 24-track machine is 50mm wide, is replayed to the performers, who add new sounds to further tracks. Lead singers and backing are added step by step. Echo is not added at this stage but is left until the final mix-down to stereo. Many studios have automated mixing facilities to fade various tracks up or down at predetermined points and to perform level control, thereby saving the need for the balancing engineer to grow extra arms! The echo mix is accomplished in a similarly automated fashion. The final stereo result may be the

157

work of no more than three performers, a producer and two energetic engineer/recordists.

Multi-tracking can be performed with just two ordinary tape recorders. A colleague of mine recorded himself singing the four parts of a barber-shop quartet by using two tape recorders. When he was satisfied with the first voice recording he played it via a mixer to his headphones and to the second recording-machine. He then sang the second voice to the first voice, both being recorded on the second machine, and so on until the quartet was complete. After echo enhancement was added to the recording he played the result to me—it was in stereo—and, allowing for the fact that all four voices were of one tonal quality, I was impressed. Multi-tracking can be performed on any system capable of playing one or more tracks while recording. It is even possible with two cassette machines, although the noise level may become rather high after the fourth re-record.

Multi-track recording-machines are superior to dual machines for multi-tracking because they do not increase the distortion level, noise or wow and flutter each time an additional track is laid. Multi-tracking with dual machines poses another problem—that of balance. After four or five re-recordings the balance may be found to be rather different from that anticipated—and practically nothing can be done to change it, except to start all over again. It is possible (just) to multi-track using several separate machines, but the problem is one of track synchronisation. It would be pure chance to achieve a synchronised start on all tracks and even when it occurred it would be most unlikely that all the tracks would remain synchronised to the end of the take.

Limiters should be used only with great care in music recording. Music (recorded) performances can be ruined by the incorrect use of electronic compressor/limiters, although very interesting effects can often be achieved with them. Music requires careful level control which should not be done electronically. Auto-level control (a permanent feature on some cassette recorders) will produce unacceptable results on music of wide dynamic range. It will result in almost constant-level recordings (lacking light and shade). If the dynamic range is to be reduced it should be done artistically, not with a sledge hammer, and the recordist must know where extra-loud sounds will occur and where prolonged low-level is likely. Therefore, unless the recordist is familiar with the piece of music to be recorded, he should obtain a score for it and know how to read it (or at least be able to follow it and look ahead a little).

To sum up, I think it is fair to say that a good music recordist is one who is both a knowledgeable technician and an artist with a discerning musical ear and a keen interest in a wide range of music. The enthusiast aiming at being a good music recordist will, no doubt, find the challenge very stimulating, for music is a field which offers great variety. But consider another field which offers an even wider variety of sounds—a field where the world's greatest singers perform, accompanied by an orchestra of many thousands of varied and beautiful sounds, conducted by the greatest conductor of them all—mother nature.

9

Recording Wildlife

The number of enthusiasts taking up wildlife sound recording is increasing year by year. Many of the old hands seem content with mono, but there are many recordists—old hands among them—who are now recording in stereo and need no convincing of the greater realism it produces. Perhaps wildlife and nature do not interest you at the moment. However, nature really does offer an enormous variety of sounds: many are easy to record, and some you may find fascinating. Once a recordist becomes enthusiastic about recording a particular subject he may also become interested in the subject itself—and nowhere is the recordist better rewarded for his efforts than in the sounds of nature.

Nature is at our doorstep—even in our houses. Some time ago I heard an amateur recording of owls in a thunderstorm. In the recording there were the calls and movements of several owls in an echoey wood; there was thunder and rain; there was suspense and drama. It was unbelievably dynamic and for the ten minutes it lasted I was spellbound by the magic of it. Superb scoops like this happen usually by chance and the recordist may achieve it only once in a lifetime; but how worthwhile it will have been, for whenever he plays that magical recording he will be transported to the original spot that he remembers so well.

It may surprise some people to realise how tame many of our wildlife subjects can be: grey squirrels may be so tame that they will sit beside one on a park bench and take food from the hand. Blackbirds, robins, chaffinches, thrushes and, of course, stock-doves will also often feed from one's hand. Like ducks on the village pond, many wildlife species have become accustomed to the presence of humans, but the same species in the wild will be up and away before one can get within twenty metres of them. It is conditioning that makes one species tolerant of another.

160

Most countries have wildlife preservation societies, and certain species may be protected by law. Anyone wilfully disturbing one of these protected creatures risks heavy penalties. Before setting out with an array of recording equipment, in the breeding season, it might be prudent for the recordist to obtain a list of the protected species for any country he is visiting. As to whether a particular species is going to be disturbed by the presence of a human depends on the nature of the locality and whether it is frequented by humans. A bird not on the protection schedule but nesting in a stream valley rarely visited by humans is much more likely to be disturbed by one passing recordist than is a scheduled species nesting on a ledge overlooking a busy tourist track. So the question of whether a species is likely to be disturbed is answered more by common sense than by the application of rigid principles.

Wildlife species can accept human presence—but not in a hurry. I have had badgers foraging a metre away from my feet, not on the first night's watch but after sitting close to the sett for several hours each night for three weeks and moving a little nearer each night. Incidentally, to aid vision I used a wide-beam lamp and a spotlight from a motor vehicle. The lamps were fitted with primary red filters, badgers being insensitive to red light. The wide beam picked out the glowing eyes of the badgers as they emerged from their holes and the spotlight was then swung round to illuminate more fully these delightful creatures.

In the Shetland Isles, in the breeding season, practically every stretch of inland water has at least one pair of red-throated divers. Pairs of these beautiful birds would occasionally settle on the loch only a few metres from where I sat at the water's edge. It was at one of these lochs that a very fine, close-perspective recording of these birds was made in stereo, using a pair of gun microphones. Recordists who rush at creatures in the wild are unlikely to get more than the alarm calls of the species they are intent on recording.

Village duck ponds are fairly accessible and ducks make most agreeable and varied quacking sounds. Anyone wishing to try making some bird recordings would do well to go to a duck pond—the larger the pond the more exciting the sounds—taking along a few food scraps. Position a microphone about

161

30cm above the water's edge and throw a handful of scraps into the water immediately in front of the microphone—then start the recording machine. The ducks will be heading towards the microphone almost before the scraps hit the water. If the recording level is controlled, the recordist cannot get anything but a first-class close-perspective recording. Other fairly easy subjects might include sheep and lambs; chickens and farmyard geese; the gulls at seaside holiday resorts; communal birds, like the rook and kittiwake; wasps, bees and flies, grasshoppers and crickets.

Wildlife creatures are not the only natural sound-makers: there are the sounds of the sea, waterfalls, gurgling and bubbling streams, and water dripping in caves. There is rain and thunder, a vast variety of wind sounds, the creaking of tree boughs in a gentle breeze and the sounds of seed pods splitting open in the heat of the sun.

The mention of heat reminds me of an occasion when I was in southern France. I took a fresh tape from the van and placed it on the recording machine in the early morning sunshine. Before long the tape was weaving about the heads and recording erratically. I tried three new reels with a similar result until it dawned on me what was happening: the sun was quickly heating and expanding the top of the previously rather cold tape, causing it to deform. Now whenever I use a machine in the full sun I cover the reels to prevent heat distortion.

Sound levels

I shall deal with mono and stereo recording techniques together, but mainly in stereo terms, in the knowledge that when the mono recordist fully realises the potential of the stereo medium he will be eager to exploit it for wildlife recording.

Ideally, the wildlife recordist should aim at producing:

(1) habitat atmosphere recordings, exhibiting no predominant sounds and no distracting noises;
(2) species recordings of medium perspective with a pleasing background;
(3) close-perspective species recording with almost neutral low-level background.

Consideration should be given to seasonal variations in the above. The distance that the microphones will be positioned from a subject for a given perspective will depend on the sound level produced by the subject, the general background sound level and the acceptance angle of the microphone system. Subject-to-background ratios are very much a matter of personal preference, but a rough guide might be: medium distance 2.5 : 1 (8dB); medium-close 5 : 1 (14dB); close 10 : 1 (20dB). Ratios higher than 10 : 1 produce a very close-perspective sound—sometimes considered clinical and unnatural.

The wildlife recordist only occasionally encounters high sound levels, such as when he accidentally stumbles on a herd of wild elephants trumpeting hysterically at close range. With loud sounds the recordist may be able to obtain good recordings with an omnidirectional microphone for mono or a pair of cardioids for stereo. But in the main the sound level averages about 50dB acoustic, which is less than 1 per cent of the level to be expected at the microphones in the concert hall. Compared with the level from a full orchestra imagine the level given at a distance of 20m by a little bird like the robin—quite an insignificant level really. It should be borne in mind that listening-level adjustments may increase low-level sounds beyond their original field levels, which will present serious noise-level problems.

An omnidirectional microphone used for species recording will need to be placed close to the subject to achieve recordings low in background atmosphere level and low in hiss level. This is not easy or, in most cases, even desirable. It usually causes the subject to flee or, in some instances, the subject may induce the recordist to depart with a degree of unbecoming haste. In placing an omnidirectional microphone for habitat atmosphere recording the same problem of disturbance arises—perhaps to a lesser degree in woodlands, due to the cover provided by trees—but on marshland and mud-flats most wildlife will be driven off by the recordist when he goes out to position the microphone.

Recording estuary sounds from the cover of a copse or hedgerow is usually unsatisfactory because, unless the recording is done at night, there is almost certain to be dominant woodland or hedgerow sounds. The distance, too, will be a further problem: the sound level from marshland subjects will be very low, necessitating high signal amplification with consequent high hiss level. If the recordist is prepared to risk driving off all wildlife by walking out to set up his microphones and laying lengths of cable back to the recording point, and if he is also prepared to wait for long periods, very good sounds may be forthcoming.

A HIDE

A small, camouflaged, portable hide will provide cover for the recordist and can be used as an alternative to walking all the way back from the microphone point to a remote recording location. The hide can be set up on a marsh and only a short length of cable need be used between the hide and the microphones. The recordist will probably have to be very patient and may well need a book (or knitting) while waiting for interesting wildlife subjects to appear. Sitting out on a marsh can provide a bonus should the recordist also be a keen photographer, for wildlife may well come near the hide and that can be as fair a reward for patient waiting as a good recording.

DISGUISE

It is possible to approach closely to subjects by using disguises which bear a close resemblance to the creatures or objects which the subject is used to seeing in its territory and which cause it no alarm. A sheep's skin would be useful, as would some form of cow or horse outfit. Photographers have used such disguises with great success—so why not sound recordists? Be careful, though, what you wear and where you wear it. A photographer was wearing a deer outfit to approach deer on a moor, when he was mistakenly identified as a deer and shot by a member of the hunt which was active that day. In boulder-strewn areas an imitation boulder could be used. It might be made from a cardboard box (of the size used to pack washing-machines) and painted the same colour as the boulders in the area being worked.

Choice of microphone

Even though the recordist may manage to get close to his subject, a close-perspective recording is not necessarily

164

achieved. With a single subject the sound source is almost a point source (a source of very small angle). Omnidirectional microphones pick up sounds from all directions equally; therefore a small-angle source uses, at best, only 1 per cent of the directivity angle of the microphone. The use of a microphone of such poor directivity efficiency means high noise levels on any recording where the subject is either not very close or not very loud. Directivity efficiency need not be confused with electrical efficiency, for two microphones of very different polar (directivity) responses may have identical electrical efficiencies and will thus produce the same output levels of signal and hiss. But, compared with an omnidirectional microphone, a highly directional microphone results in far less ambient sound pick-up from the surroundings. This is of great significance as, for a comparable result, it enables the recordist to record at greater distances from the subject.

A good cardioid microphone is capable of attenuating sounds to the rear of itself by 20dB (a voltage ratio of 10:1); when used to record estuary sounds from the edge of a copse, the unwanted woodland sounds would be significantly reduced. In considering low frequencies, ie traffic rumble, remember that some microphones described as cardioid may be practically omnidirectional at frequencies below 200Hz and would give little reduction in traffic rumble from any direction.

Of all the various fields of recording, wildlife is probably the most demanding. There are always so many variables to consider and so many problems to overcome—usually by the recordist on his own and often a long way from civilisation. For me, though, it is also the most challenging, the most absorbing and the most enjoyable field of recording. One of the problems is to obtain recordings with good signal-to-noise ratios. Noise is the great bugbear, particularly electronic noise (hiss) and, owing to the high amount of signal amplification necessary to obtain a good recorded signal level, it is the fortunate recordist who obtains a signal-to-noise ratio of 40dB on atmospheres. With species recordings the signal-to-noise ratio must be higher than 40dB because, with much less background sound to mask it, the hiss would be clearly heard.

Dynamic (moving-coil) microphones are the most suitable

for outdoor use; they are robust and reliable, not affected by temperature or humidity variations and are not likely to be affected by strong magnetic fields when out in the open. When selecting microphones for wildlife recording take care to choose only those with a high sensitivity (0.2mV/μbar or higher) and a good high-frequency response. Capacitor microphones are prone to humidity and dew-point problems so cannot be recommended for reliable service in the variable conditions of outdoor working.

The parabolic reflector has been associated with wildlife recording for years, deriving popularity from its ability to concentrate a very small angle of sound on to a microphone situated at its point of focus. The parabolic reflector's performance and shortcomings are well known: the forward low-frequency performance is limited by the diameter of the reflector and low frequencies tend to be picked up equally from all directions when not controlled by a very large reflector, ie around 3m diameter; the middle frequencies (500Hz–3kHz) are of fairly wide angle; the high frequencies very narrow—assuming that the microphone (which faces into the dish of the reflector) is set on-axis and at the correct point of focus. The forward gain of a reflector is low at low frequencies, high at middle and high frequencies but, as just pointed out, the pick-up angle at middle frequencies is fairly wide—much wider than at high frequencies (the polar response of a reflector is frequency-dependent). Therefore the *quantity* of sound picked up at middle frequencies is greater than at high frequencies. The consequence of this narrow-angle frequency-dependent polar response is that when the call of, say, a bird is received a little off-axis it will have a different tonal quality compared with the same call on-axis. Hence, when recording a small group of birds, the tonal quality of any single bird's call may vary significantly as it moves around in the group.

The parabolic reflector is not a device which can be said to produce high-quality results, particularly when used to record atmospheres. When used to record individual subjects or tightly packed groups, the sound quality of a reflector is quite acceptable on a wide range of wildlife subjects. The one great advantage of the reflector over other microphone systems is acoustic amplification; the signal-to-noise ratio of an on-axis reflector-received signal, albeit a narrow-angle one, should be very much better, both electrically and acoustically, than with any other microphone system for a given subject-to-microphone distance.

An omnidirectional microphone may be used in a reflector, but a cardioid microphone is recommended. There are three reasons for this: first, when the microphone is used separated from the reflector it will be more useful if it has a good directional response. Second, when a cardioid microphone is used, the reflector combination is less sensitive to sounds from the sides. Third, the directional sensitivity of a reflector falls sharply at a certain off-axis angle. At a frequency dependent on the distance between the microphone and the reflector surface the reflected sound signal at the microphone will be equal in level but opposite in phase to the direct signal received by the microphone and as a result there will be cancellations—there will be a dip in the frequency response. With a reflector of 20cm focal length this dip will occur right in the middle of the frequency range of the reflector, at 850Hz–1000Hz. When an omnidirectional microphone is used in a reflector, cancellations can occur on signal angles as small as 10° off-axis. The sensitivity at the rear of a cardioid microphone is very low and, when one is mounted in a reflector, its dead zone is facing the direction from which the reflector receives its signals. Therefore any cancellations which do occur will be to sounds a long way off-axis and at levels so low as to be insignificant.

The reflector is very sensitive to handling noises—even with small units it can be a serious problem. Flexible antivibration mountings for securing the microphone to the reflector help to reduce noise, but handling or contact with grasses and twigs causes the whole reflector shell to vibrate. Soft rubber or felt secured to the back and edges of a reflector will reduce the impact, and thus the noise, of small objects touching its back and edges but such materials are absorbent and become very heavy when wet.

The recordist working in mono should be able to carry a small reflector of about 60cm diameter in one hand, a cassette or open-reel recording machine across one shoulder and an accessories bag across the other. The bag should contain a plastic raincoat and overtrousers (a prerequisite of working outdoors in Britain), spare tapes and batteries, a pencil and notebook, a species identification book, headphones and binoculars (where these are not already carried around the recordist's neck), a screwdriver and a pair of pliers, a good length of strong rope, a torch, a first-aid kit and a bag of buns. A length of rope is essential for the safety of a wildlife recordist because, when clambering down steep rocky slopes, it is far safer for him and his equipment if a rope is secured at the top of the slope to assist his descent.

The actual time that a wildlife recordist spends recording may be very short. Usually there is need to explore an area fully to discover what wildlife it harbours. As much as 85 per cent of a recordist's effort can be spent on reconnaissance and in areas of difficult terrain it becomes expedient sometimes to reconnoitre without taking a load of equipment. For recordists who travel light a well-known law operates. It suggests that the best material is available always when one least expects it and is least prepared for it. On some days the recordist may find nothing worth recording, although he is ready to do so on the instant. Another day may provide such a variety and abundance of material that he could go on recording for hours were it not for the tape or batteries running out.

The mono recordist equipped with a reflector really has two microphones in one. He has a narrow-angle long-range device (which may be of considerable size) and he has a small cardioid microphone (by separating it from the reflector). For recording a bird singing in a bush, for example, a reflector-mounted cardioid at 20m distance may achieve a result comparable to that obtained at 3m when the microphone is used on its own. The recordist is unlikely to be able to approach much closer than about 10m without disturbing the bird. First, therefore, a reflector recording should be obtained from a reasonable distance. Then, if the bird is observed to return frequently to the same bush, a recording of different perspective and better quality may be possible with the cardioid microphone removed from the reflector. It can be positioned in the bush and the recordist can retire to convenient cover to await the return of the bird.

Of course, the next bird which visits the bush may not necessarily be the one expected and this is where the recordist will find his binoculars invaluable, for while the bird is being recorded a correct identification of species can be made. When relying on a bird's oral renderings for identification it is as well to realise that many birds are good mimics and may well deliver song phrases belonging to other species; therefore a visual identification becomes imperative.

One of my colleagues tells of an occasion when he set a microphone in a bush and retired to await the arrival of a bird of the species he had decided to record. Shortly after reaching the recording point and putting on his headphones he heard human voices approaching and was surprised when a courting couple took up temporary residence under the very bush containing the microphone and, after some preliminary discussion, proceeded to a variety of fascinating activities. My friend was so enthralled by the drama that he completely overlooked the recording opportunity.

Reflectors will be mentioned again later in connection with their use for stereo recording, but now we shall continue with our examination of more manageable microphones.

OMNIDIRECTIONAL MICROPHONES

Omnidirectional microphones allow for an all-round pick-up of sound and will be considered for stereo recording. When two omnidirectional microphones are spaced 15m apart in, say, woodland habitat, the sounds which arrive at both microphones at the same instant will be from a position 90° from the centre of the line between the two microphones, and these sounds will be reproduced in the centre of the sound-stage. One of the factors enhancing woodland sounds, and which also greatly assists our direction-sensing, is the reflective properties of the location. Our appreciation of direction and space is governed by the daily reprogramming we receive and this is related to the response variations of, and the distance between, our ears. By using coincident microphones for loudspeaker stereo, and by using microphones spaced about 15cm apart for binaural listening through headphones, we are presented with information which our ears can interpret. By increasing the distance between two microphones, the time and phase relationship of received signals will

169

be outside the limits which our ears normally experience and confusion will result.

A similar set of circumstances results when one of the loud-speakers in a stereo system is positioned at a greater distance from the listener than the other, the listening angle (70°) being maintained and the level differences being compensated for. The phase of the signals from the more distant loudspeaker will lag the signal phase from the nearer one and produce rather unpleasant listening.

In woodlands, the omnidirectional microphones need to be spaced at least 20m apart to produce any worthwhile stereo effect, and signals from any angles either side of the centre 90° position will arrive at each microphone at different instances and be different in phase. Now, consider sounds arriving on a line with the two microphones: the second microphone will receive its signal some 20m later than the first, ie 58ms later, and this will produce a pronounced flutter echo.

The spaced microphone system exhibits variations in volume on near subjects. For example, if a subject passes nearer to the microphones than their distance apart there will be a volume dip at the midway point. The subject volume then returns to full level as it passes the second microphone and dips away again, accompanied by flutter echoes, as it recedes from the microphones. Another drawback of wide spacing is that it produces blocks of localised mono sound at both sides of the sound-stage and, usually, there is a noticeable hole in the middle. The movements of subjects are also poorly conveyed and the phasing is generally unpleasant. Several stereo records of wildlife subjects have been produced over the years and it is not too difficult to pick out the ones which used spaced microphones to record the original material, for they exhibit most of the problems just mentioned.

A recordist who has two omnidirectional microphones and wishes to use them for stereo recording should fix the microphones one on either side of a piece of felt-covered board about 30cm square. The microphones should be spaced about 5cm from the board, with their heads just forward of the centre. If the rear pick-up is reduced by fixing a large, heavily felted board at the back, the system may produce reasonable results. This device uses physical masking to achieve a form of directional response; the microphone on the right is shielded from picking up sounds on the extreme left, and vice versa, while both are shielded from rear pick-up. The directional response is frequency-dependent and the system cannot be expected to perform as well as proper directional microphones. However, the results will be better than those from spaced microphones and, in fact, can be rather impressive through headphones.

CARDIOID AND SUPER-CARDIOID MICROPHONES

The mono recordist will find that a super-cardioid microphone provides a very good forward angle for the desired sounds and a very good dead zone at the rear to attenuate unwanted sounds. This characteristic can best be utilised by positioning the microphone between the desired and the undesired sounds, and a good super-cardioid should then reject 90 per cent of the undesired sounds (unless some of the rearward sounds are reflected back by objects in front of the microphones). The recordist should also be able to obtain recordings with less general ambience by using a super-cardioid in preference to an omnidirectional microphone. Cardioids are not quite as directional as super-cardioids and they pick up a little more ambience, but the dead zone at the rear should be extremely beneficial in diminishing undesired sounds.

For stereo wildlife recording cardioid and super-cardioid microphones perform extremely well. The inter-axis angle will require careful consideration because, if the angle is too wide, the sound-stage will exhibit pools of mono from each side. For atmosphere recording, where sounds issue from a 360° angle, inter-axis angles will be roughly 50° for cardioid and 60° for super-cardioid. In fact, once the angle is set to produce good stereo on atmosphere it can remain more or less unchanged for most outside work: even a close-perspective recording of a subject will have some atmosphere and, if the angle has been set for good stereo on atmosphere alone, it will also be correct for the atmosphere accompanying the subject.

Inter-axis angles may be increased when recording loud sounds from limited angles in locations low in general background. For example, when recording a noisy gaggle of geese on an otherwise quiet backwater, the angle may usefully be increased to 90° for cardioids and 75° for super-cardioids. The best systems to use for a close-perspective recording of the geese are the three- and the four-microphone super-stereo systems described in Chapter 8.

For recording 'at the nest', rather small microphones are required and I find that the AKG C451 capacitors with angle-joints fit the bill nicely. The head capsule can be angled at up to 90° from the body of the microphone, so that the bodies of two microphones (for stereo working) can be positioned closely together in the vertical, while the capsules cross one over the other (coincident) at 50°. The whole assembly is enclosed in a special windshield of 13cm diameter and is small enough to be used for close working. When in-line microphones are used the diameter of the windshield would need to be at least the equivalent of twice the length of one of the microphones. Microphones are rather long but they contain only a cable connector, possibly a switch and sometimes a small transformer.

The main purpose of the long body is to provide a handle for the user; however, the wildlife recordist quickly learns not to hand-hold a microphone when recording, for it only results in extraneous noises: better to fix it to a post or even a microphone stand. The microphone's active part—the insert element—is only a fraction of the total length and it is sometimes possible to obtain replacement elements. I recently obtained four cardioid inserts, as used in a popular dynamic microphone, and with these a compact, four-microphone system was constructed. The wildlife recordist does not need expensive polished microphone handles, for he is not necessarily a showman, and inserts may provide an acceptable substitute. They are small in size, light in weight, easy to install and about one sixth of the cost of the complete microphone. Capacitor microphone elements will not perform on their own: they require an amplifier of high impedance and a 50V polarising supply. The electret type requires only the high-impedance amplifier.

HYPER-CARDIOID MICROPHONES

The forward angle of a hyper-cardioid is narrower than that of a super-cardioid. Its value to the wildlife recordist comes from the much sharper and more selective front zone and the good rejection zone centred at about 130° all round. It is possible to use a hyper-cardioid in locations troubled by single-direction noises and obtain clean recordings. Rejection of, for instance, a farm tractor in a nearby field may best be achieved by positioning the microphone between the desired sounds and the tractor noise, and pointing it upwards at an angle of 50° to the plane of the land. The tractor will then be in the maximum rejection zone of the microphone and the noise will be reduced to perhaps 10 per cent of that which would be picked up on-axis.

When an aircraft is spotted approaching the recording area, the microphone can be turned to present its dead zone to the passing aircraft to reduce the noise pick-up. The noise angle will always lag the visual angle where aircraft are concerned and this will need to be allowed for. Hyper-cardioids are of greatest benefit where they can be positioned (on a stand) pointing either upwards or downwards at about 50° to the horizontal to reject horizontal noise from a particular direction. Fortunately, both the mono and the stereo recordist can take advantage of this. It is the only stereo microphone system which can be arranged to reject horizontal noise equally on both stereo channels, and this can be achieved by pointing the system upwards (or downwards) at an angle which causes the maximum rejection of both microphones to coincide with the noise direction (this will be at 180° to the common microphone axis). The inclination of the system will need to be carefully angled to achieve maximum rejection but the results obtained may justify the extra care and effort.

With the increasing amount of man-made noise around us good rejection characteristics are becoming more important than good forward acceptance in microphones, and a hyper-cardioid provides good rejection and good acceptance. Its characteristics make it ideally suited to medium- and close-perspective species recording.

FIGURE-OF-EIGHT MICROPHONES

The mono recordist will find the figure-of-eight microphone a most versatile instrument. It can be used successfully in conditions which preclude the use of any other microphone, for it is

173

better than a cardioid and better even than a hyper-cardioid at noise rejection. It can be used in the horizontal mode: under a tree to capture the sweet renderings of an elusive bird, or over a badger hole to record the emergence and romps of the creatures, where the horizontal noises from motorways and farm machines may be tuned out to an astonishing degree. In the normal (vertical) operating mode barking dogs at the nearby kennels can be tuned out by the careful direction of the microphone. It is possible to achieve a rejection of up to 40dB (1 per cent of the on-axis level) with a good figure-of-eight microphone. In fact, the noise of a farm implement in an adjoining field can be reduced to a level equal, and perhaps below, the general atmosphere level on a recording.

Unfortunately, the stereo recordist cannot take full advantage of this very convenient characteristic because the dead zones of the two microphones will operate in different directions, so that when in the normal (vertical) operating mode one microphone is set to reject an interfering noise, the other will pick it up. To reject equally on both channels the noise must arrive to the plane of the ribbon of *both* microphones, which means that the microphones must be used in the semi-horizontal mode to reject horizontal noise, and then the rejection is purely bidirectional (end on).

One problem to overcome when figure-of-eights are used for wildlife recording in stereo is phase. It is not a serious problem where the forward angle of sound is limited to about 120°, but when there are sounds from all round, the sounds from the sides cause very unpleasant out-of-phase signals in stereo, and in mono they cancel. Coincident figure-of-eight microphones cannot be used in all-round sound locations, unfortunately, so a side-by-side, slightly spaced mounting has to be adopted.

From Fig 33 it can be seen that the sound from the side is directed to the front of microphone B and to the rear of microphone A. Thus the phasing will be B-positive and A-negative, which is unacceptable to the stereo listener, and the mono result will be zero. However, as far as the middle and upper frequencies are concerned, this problem can be overcome by fixing a felt-covered acoustic screen (about 23cm square) between the two microphones. The side sounds reaching the rear of the micro-

174

phones will be greatly attenuated by this screen but it will have very little effect on the sounds coming from the front and from behind the microphones. The improved version is capable of very good stereo, although I personally consider 360° of field sound presents rather a crowded scene when condensed to a 70° sound-stage.

NARROW-BEAM GUN MICROPHONES

To the wildlife recordist a gun microphone is a very worthwhile instrument in that from a whole field of sound it favours only a small angle—a cone of about 60°. Sounds outside this accept-

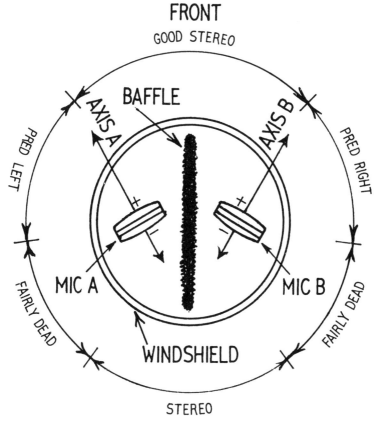

Fig 33 A pair of figure-of-eight microphones with a centre baffle

175

ance angle are substantially reduced but the microphone has no rejection zone and is not suitable for recording in conditions of high extraneous noise. The excellent selection characteristics of the gun microphone enables the recordist to achieve good species recordings and habitat atmosphere recordings, particularly on estuaries, marshes and moors. In these locations most of the surrounding sounds are indigenous and do not necessarily need to be rejected and, in any case, being wide of the acceptance angle of the microphone they will be heard only at a much reduced level.

When omnidirectional microphones are used in woodlands and on marshes and estuaries there is often too much background roar, and the species sound too crowded, with masses of birds all at about the same perspective. Recordings taken on gun microphones have less atmospheric roar and are less crowded because the species not in the acceptance angle sound much more distant, thus improving overall perspective. It is more necessary to achieve clear, uncrowded recordings with good perspective when working in mono than when working in other systems because the sounds in mono are for reproduction through a single loudspeaker.

Gun microphones for stereo wildlife recording should be crossed at about the capsule area and angled at 28°, as described in detail in Chapter 5. The acceptance angle for stereo then becomes about 70°, which is the same as the sound-stage. The stereo results will be very good over the 70° angle but there will be considerable phasing of multiple-frequency sounds moving about this angle, owing to the spacing of the microphones. Phasing is particularly poor when a pair of gun microphones is positioned with the tail ends of their windshields just touching, or at greater separation. This is well demonstrated on recordings of human voices moving about the acceptance angle when the recordings are replayed in mono: the sound quality of the voices will vary according to their positions in front of the microphones. However, for the majority of wildlife sounds, a pair of gun microphones produces a very good stereo performance when correctly mounted.

I have used gun microphones in separate individual windshields and in a special one-piece windshield enclosing both microphones. The one-piece windshield not only provides a more manageable system but it produces a more cohesive stereo wind effect. The use of separate windshields produces the very separated left and right wind buffeting noise so characteristic of individual spaced microphones.

PARABOLIC REFLECTORS FOR STEREO

As stereo consists basically of two channels of interrelated monophonic sound, it would at first appear that long-range recording in stereo requires two parabolic reflectors. However, this is not so. In 1971 I was engaged in a series of tests with reflectors. In the first test two 914mm diameter reflectors were positioned closely side by side and angled in stages from 0° to 15° outwards. The position of the microphone in each reflector was also varied from long to near focus and different microphones were tried. The results of the tests were as follows: there was no stereo in the range 500Hz–2kHz; at 2kHz and above there was very poor stereo and if the angle between the reflectors was increased to more than 5° a hole occurred in the middle of the sound-stage.

The fact that there was no stereo at all at low frequencies was no surprise because, due to the very wide pick-up angle of the reflector at low frequencies, the sound intensity arriving at each microphone could not be anything other than identical. The hole in the middle of the sound-stage at high frequencies resulted because reflectors have a very narrow acceptance angle of about 5°, so that when they were angled at 15° there was a 10° uncovered (dead) angle between them. The results of this experiment show that two reflectors placed closely side by side give an unacceptable stereo performance.

The second experiment used the same set-up but the distance between the reflectors was increased in stages up to 30 paces apart and the axes varied from straight ahead to 30° crossed.

After many hours of experimentation I came to the conclusion that the use of two parabolic reflectors for stereo produces poor results. Considering the difficulty of manhandling one 914mm reflector I was not disappointed for, as luck would have it, my colleague John F. Burton and I had recently heard that a Swedish wildlife recordist/naturalist, Dr Sten Wahlström, had subdivided a standard reflector with a piece of board and mounted a microphone on both sides of the division. As he seemed rather pleased with the results of this dividing board, we decided to convert one of our reflectors to assess the stereo potential of the system. A 6.5mm thick plywood acoustic baffle was contoured and fitted across the reflector dish splitting it, effectively, into two vertical half-sections. After further experimentation, the microphones (cardioids) were positioned with their heads about 25mm outside the long-distance focal point and inclined slightly inwards to face the board. Complete felting of the board and many other variations were tried, but the best results were obtained with the board plain and reflective.

Stereophonically, the background was reasonably good and movement of subjects 10° either side of the axis of the reflector gave substantial stereophonic movement in the sound-stage. The reflector is a very narrow-angle device and the pick-up intensity falls off rapidly beyond 10° off-axis. There were phase problems with sounds coming from 10° or so vertically above and to the side of the reflector axis. It was noted that where 'calls' contained both low and high frequencies the reproduced low frequencies did not come from the same point as did the high frequencies, ie when the subject called left (high frequencies) the low-frequency part of the call swung to the right on subject positions 5° or more left of centre, and similarly on the other side. As relatively few wildlife subjects for which a reflector would be used have vocalisations containing low frequencies, the low-frequency positional shift is of little concern. The divided reflector, therefore, is the only instrument available at the moment for long-range recording in stereo. Indeed, the results are very worthwhile once the correct position for the microphones has been found.

As far as I can assess, the reflector works in a most unconventional manner. For instance, the microphone on the right receives its signal, in a stereo sense, from the left and therefore becomes the left channel; similarly, the microphone on the left becomes the right channel. The atmosphere is picked up, in the main, by the microphones directly and the reflector board only acts as a stereo channel separator; thus the atmosphere pick-up is directionally normal. The atmosphere on its own does tend to

178

have a hole in the middle but this is not too obvious when the centre of the sound-stage is well filled by subject material collected by the reflector. The low-frequency shift is probably due to the direct pick-up on the microphones being equal to, or greater in strength than, the reflected signal, because the efficiency of the reflector is very poor at low frequencies.

Fig 34 shows the right-side section (stereo left for reflected sounds) of a divided reflector. An on-axis 0° signal is directed to the centre-board focal point, which in turn reflects it straight into the microphone. The microphone on the other side of the board will receive exactly the same signal at exactly the same time and hence the reproduced sound will be stereo centre. When a signal comes from 5° right it is concentrated on a point inside the focus and is then reflected away so that most of its energy just misses the microphone. A signal from 5° left is reflected just beyond the focal point but on its way it hits the microphone directly. Hence, between 0° and 5° left, the microphone on the right (stereo left) receives the full signal strength and between 0° and 5° right it receives a diminishing signal level. Therefore, just as with any coincident pair of microphones, the stereo effect is achieved as a result of the diminishing strength of signals to one channel or the other.

The shape and contour of a reflector affect its stereo performance. The best dish shape is a shallow one, ie where the focal point is outside the dish. Both top and bottom of a reflector tend to confuse the formation of good stereo images. To overcome this problem I built an improved version in glass fibre and, as well as giving better stereo, it assists transportation in the field, for it serves the dual functions of reflector and wheelbarrow.

Another reflector under consideration will be parabolic only on the horizontal plane, the vertical plane being straight. The idea is that increased efficiency may be achieved by using up to six microphones spaced at intervals up the vertical focal plane, ie six on either side of a centre board. A limiting factor would be the cost of all the microphone units required, although inserts would be satisfactory and would be contained in a one-piece windshield.

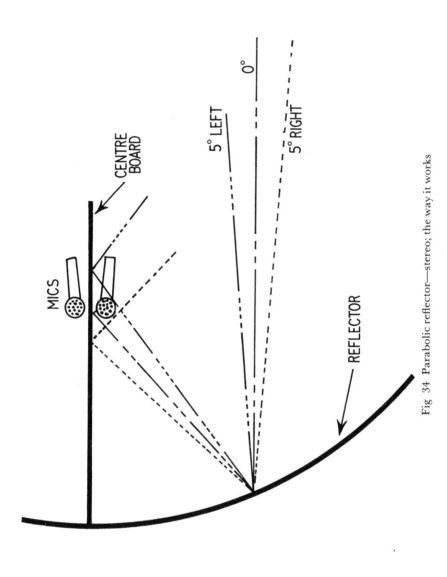

CENTRE BOARD

MICS

5° LEFT

0°

5° RIGHT

REFLECTOR

Fig 34 Parabolic reflector—stereo; the way it works

Balance of subject and background sounds

The art of achieving good balance on subject and background and also on subject-to-background ratios will come with practice and a keen ear. When attempting a recording, an experienced recordist will quickly decide which microphones to use. If the subject is likely to depart when approached, the stereo reflector might initially be used, for an inferior recording is always better than none at all. After that, one might possibly try a pair of gun microphones for two different perspectives. Then another system could be attempted for close perspective. The recordist should endeavour to record subjects in at least three perspectives: medium-distant, medium-close and close, all with correctly balanced backgrounds. If the subject is not positioned dead centre this is not too serious as long as the background is balanced.

Once a microphone system has been placed at a distance, it is impossible to change its direction without some means of changing it remotely. It has on many occasions been necessary for me to change the direction of microphones to obtain a better balance when a subject under study has changed its position. To overcome this problem I adapted a small motor and gearbox unit to fit on to a microphone stand, with the microphone windshield assembly fixed to the slow-motion output shaft. Power is supplied along the screens of the microphone cables and fly-lead connectors are used at the ends to connect the cable screens to the supply and to the motor. The drive is reversible, reasonably quiet and slow enough not to cause concern to subjects quite close to it.

Another unit which might be extremely useful (one I have not yet developed) takes the form of a radio-controlled, tracked vehicle which could creep up on a subject without causing it concern; it could be disguised to look like a small bush or an anthill. Radio-controlled model tanks and other vehicles are readily available. Although the idea of using such vehicles to approach one's subject may seem comical, it is by no means outside the realms of feasibility and before very long it may become quite a normal method by which the shrewd wildlife recordist contrives to get his microphones where he wants them.

The positioning of microphones can affect the overall balance

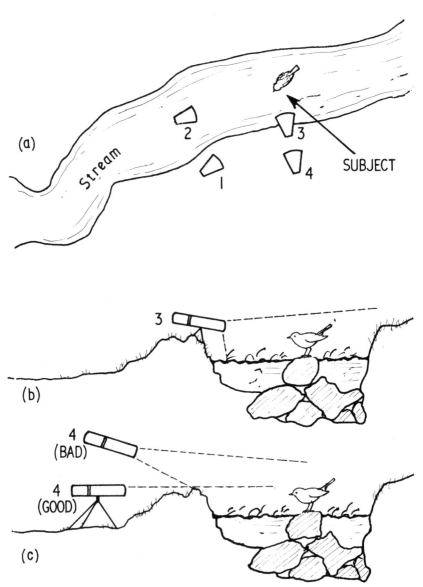

Fig 35 Microphone positions for recording subjects on water courses (a)
stream, subject and four different microphone positions. Only position 4 is recommended (b) microphone too close to stream noise (c)
stream bank provides acoustic noise shield in position 4 (low)

of a recording. Let us consider recording beside a noisy stream. If the microphones are placed to one side of the stream and directed to face up or down the stream, the result will be a heavy noise on one side of the sound-stage.

In Fig 35 (a) position 1 will result in high-volume stream noise on the left and low-volume country sounds on the right. Position 2 will result in excessive and close stream noise, though possibly balanced. Position 3 will also result in excessive and very close stream noise. Position 4 is capable of good results because the stream noise will be very much reduced—probably by 20dB or more when compared with positions 2 and 3. The practised mono recordist will know that, when using a reflector, stream and seawash noise is least audible when the reflector is positioned close to the ground, avoiding the direct line-of-sight sound-waves from the water.

The effectiveness of shielding can be experienced at the seaside or river-side, particularly where there are banks built up above the level of the surrounding land. Standing up on the bank the full noise from the water will be heard; then, by lowering one's head to the level of the top of the bank an appreciable noise reduction will be noticed. On moving a small distance back from the bank, maintaining the same level, a further reduction in noise will be evident. Usually, the further one retreats from the bank the less will be the noise. Fallen trees, old oil drums, boats and any mound of material can be used to provide a noise barrier.

STREAM SILENCING

Where an aquatic subject can be relied upon to return to the same location following initial disturbance, or where a nearby stream causes problems when recording non-aquatic subjects, an additional method of water-noise reduction may be practised—stream silencing. Usually, all that is necessary is to locate the areas of noise and rearrange boulders and stones to smooth out the water flow at these points. Planks of wood can be used to convert a noisy waterfall into a very much quieter water-shoot.

A large piece of thick canvas can also be employed at waterfalls. The canvas should be secured at the top of the waterfall,

183

above the water level, and allowed to drape over the fall. It should be long enough to reach the stream below, where boulders hold it in place. The water should run down the inside of the canvas and so reduce splash. The sheet itself acts as a noise shield too. The same large canvas sheet could also be used *over* areas of stream noise. It could be secured above the water level and allowed to float on the water running downstream, effectively silencing it.

Noises overhead, like leaf rustle from trees, can be reduced by placing a piece of board (preferably felted) above the microphone assembly, as a baffle. To reduce ambient noise when recording insects in grass an acoustic box, usually of transparent material and with a microphone fitted inside, is placed over the insects.

The means by which noise can be reduced, particularly frequencies above 500Hz, are innumerable and I hope that the few examples given here will encourage recordists to experiment with these and other ways of reducing noise. It should hardly be necessary to mention that any stream or waterfall which has had 'silencers' applied for recording purposes should be restored to a reasonable state of normality after the recording session.

Windshielding

Wind is a mass of air in motion which may cause objects in its path to move and produce noise. It also causes blasting noises on microphones. There is little one can do about noises caused by wind blowing through trees, except to wait until the wind drops. There is, however, a lot one can do to prevent wind blowing directly into microphones. The susceptibility of microphones to wind noise varies according to type and there seems to be a common belief that an omnidirectional microphone is less affected by wind than any other type but, as will be shown, this is not necessarily true.

Wind noise on microphones can be caused by air-pressure fluctuations and by air velocity. All microphones are affected by air velocity, while only certain types are affected by pressure fluctuations. Even in locally calm conditions, pressure fluctuations may be present around chimney stacks, trees, earth mounds, between and behind buildings—in fact, in nearly every place a recordist might shelter from air velocity (wind). The point is that, although there may be zero wind velocity in such sheltered places, there can be very large pressure fluctuations at frequencies up to 15Hz. A pressure-sensitive microphone, like an omnidirectional one, may produce very large output signals at these low frequencies and, although these will not be directly audible, intermodulation products may well cause severe distortion on a recording. Omnidirectional microphones may, therefore, prove to be the least suitable for outdoor work.

The ribbon figure-of-eight instrument is very much affected by air velocity; in fact, a small gust of wind may be enough to fracture the delicate ribbon and for this reason it requires a very good windshield for outside use. The ribbon, however, is totally unaffected by pressure fluctuations because, at the very low frequencies which are caused by obstructions, the pressure will be identical on both sides of the ribbon.

Most cardioid and hyper-cardioid microphones have reasonably free air access paths to the rear of the diaphragms and they are not badly affected by low-frequency pressure variations.

Wind velocity noise, as heard on a tape-recording, varies from filterable rumble, with all the desired signals quite clear, to a continuous disembodied tearing sound. The tearing noise is the amplified sound of wind eddies round a small obstruction: the smaller the obstruction the higher the frequencies of the eddies. Unlike the previous example, in which there were low-frequency eddies round large obstructions, we see here that the presence of small obstructions in an air flow causes high-frequency eddies of a complex nature, producing audible noise over a wide range of frequencies.

A microphone with a ball-type wire mesh head will produce high-frequency noise caused by eddies round each filament of the mesh, as well as lower-frequency noise caused by obstruction of the microphone as a whole. There may also be blasting as a result of wind blowing through the mesh. Many microphones incorporate pop filters to reduce the blast of air produced by close speaking. Pop filters are air velocity-reducing screens fitted between the mesh head and the diaphragm of a microphone. They are not musically selective as the term might suggest! To draw a parallel, consider a wind blowing through an

185

open window at which we are standing: we are subjected to the full velocity of that wind, but when an acoustically transparent cloth is placed across the window opening the wind velocity is reduced without reducing the transmission of sound through the opening. The shape of the fibres in the cloth affects its efficiency as a velocity reducer, and the sound transmission efficiency is related to the spacing of the fibres. The higher efficiency of flattened fibres over the more common round-section fibres is most marked.

The purpose of an external windshield is to provide a smooth air flow over the microphone assembly without degradation of sound quality. The simplest type of windshield, the gag or muff, is made from a porous material such as open-cell rubber or plastic foam. It fits directly over the head of a microphone and provides some protection against close-voice 'p'-blasting and small air currents. The efficiency of a muff is limited by its size, which unfortunately has to be small to avoid degradation of frequency and polar responses.

The effectiveness of any windshield depends largely upon the signal level to be recorded. Low levels require more amplification, applied not only to the signal but also to any wind noise which may be present. The wildlife recordist requires very efficient windshields, perhaps in conjunction with simple muffs, because generally the signal levels are relatively low and the wind speeds are relatively high. The aim in good windshielding is to provide a total enclosure for the microphones in which the air velocity is almost zero, irrespective of the wind velocity outside the enclosure. For minimum air turbulence the best shape is a sphere, and one or two manufacturers have produced large, spherical windshields. A cylindrical shape is also very good and is very much easier to construct.

A very simple enclosure can be made from a cylinder of strong wire mesh, with either a mesh or a solid top and bottom, covered with 8mm plastic foam. The microphones should be mounted with their heads as near to the centre of the enclosure as possible and pointing so as to pick up the sound through the side. This arrangement can provide a useful measure of protection, but it must be remembered that plastic foam of any reasonable thickness attenuates high frequencies—especially when it gets damp—and it also damages very easily.

The advantage of using a high-durability fabric with excellent sound transmission and wind resistance characteristics is obvious. A fabric meeting these requirements is 'Tygan', manufactured in Britain by Fothergill and Harvey Ltd, Lancashire. I found the cloth easy to machine-sew and easy to glue with contact adhesive. I have used windshields covered with this fabric for a number of years and they are as good now as when they were first used.

In designing windshields my aim has been to produce units which would provide a high degree of protection, even in gale-force winds. By adopting a 'tent-within-a-tent' technique I believe this aim has been achieved.

Fig 36 shows the basic details of one of my designs, in which two separate 'Tygan'-covered frames of 'Weldmesh' are used. The frames are at least 15mm apart so that any wind which penetrates the first barrier is encouraged to travel around the space between the two barriers, rather than across the inner section containing the microphones. The fit between the frames and the supporting glass-fibre ring, and between the ring and the base plate, must be good in order to prevent direct wind entry to the centre. The larger the frames and the greater the space between them the better is the efficiency of the unit. The heads of the microphones should be positioned as near the centre of the inner cage as possible.

A large twin-barrier windshield of this design has been used in gale-force winds at the top of a cliff and with capacitor microphones (normally vulnerable to wind noise), with absolutely no wind rumble. It must be pointed out, in fairness, that a very efficient electronic rumble filter was also used, and this may have been a contributory factor in achieving such unmarred results in the rather harsh conditions.

I have a number of windshields: a small-diameter double-barrier windshield designed to accept a pair of AKG C451 microphones with angled head-joints (and in extreme conditions a plastic-foam pull-over muff is used to provide additional protection); a slightly larger one containing four cardioid inserts; a large double-barrier windshield for a cluster of four Beyer M88 microphones; and finally, one for a pair of gun micro-

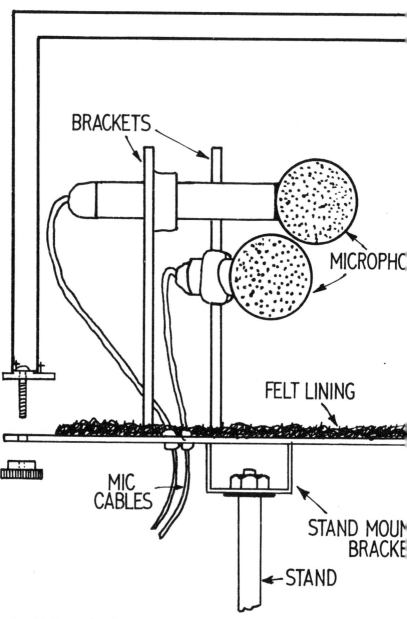

BRACKETS

MICROPHO

FELT LINING

MIC
CABLES

STAND MOUN
BRACKE

STAND

Fig 36 Double-barrier windshield. Constructional details showing correct
position of microphones

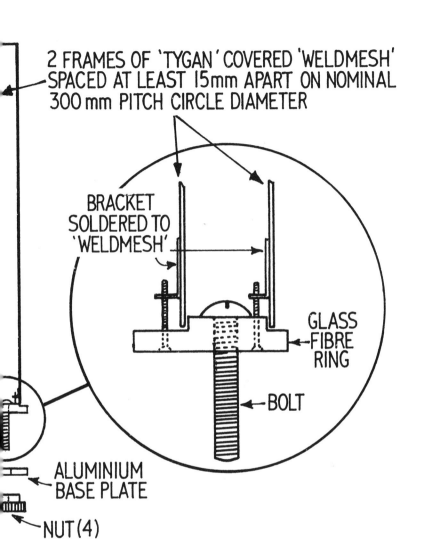

2 FRAMES OF 'TYGAN' COVERED 'WELDMESH' SPACED AT LEAST 15mm APART ON NOMINAL 300 mm PITCH CIRCLE DIAMETER

BRACKET SOLDERED TO 'WELDMESH'

GLASS FIBRE RING

BOLT

ALUMINIUM BASE PLATE

NUT (4)

phones. Because of the rather odd 'banjo-case' shape of this last one, it has not been possible to construct it on the double-barrier principle. To provide additional protection, foam-plastic sleeves are fitted over each microphone, though these are probably less effective than would be a double-barrier windshield.

It is possible to make trouble-free recordings in very high winds if precautions additional to good microphone windshielding are taken. A large cloth windbreak, similar to the type used by sun worshippers on the beach, is a useful asset. It will need to be about 1m high and 2m wide (or wider if it is also to provide cover for the recordist) and may need careful roping down to prevent the fabric flapping in strong winds. The windbreak should be arranged at about 60° to the horizontal and sloping with the wind, thus allowing an easy up-and-over airflow at the front and a relatively calm zone behind. The windbreak may often be placed between subject and microphone, and therefore must be acoustically transparent. 'Tygan' and similar fabrics are suitable but avoid cloths which whistle in a strong wind.

Even when every mechanical precaution has been taken against wind rumbles it is more than likely that some low-frequency noise will get through in windy conditions. Rumbles on recordings are very distracting and it is worthwhile using every possible means to eliminate them. Large-amplitude audible and subsonic signals can cause instability, distortion and noise within the electronics of a tape recorder; as frequencies below 100Hz are rare in wildlife recording, low frequencies should be cut before they cause trouble. Fortunately, with electronics, this is a relatively simple matter. A suitable response for microphone amplifiers designed for wildlife recording is 150Hz–15kHz±1dB, with a switchable high-frequency lift of +4dB at 10kHz and 6dB/octave. The response beyond 20kHz should be severely attenuated, to be free from radio-frequency pick-up. The bass frequencies should be cut at 125Hz by at least 24dB/octave, so that a 50Hz signal will be reduced to $\frac{1}{20}$, a 25Hz signal to approximately $\frac{1}{200}$, and a subsonic signal below 15Hz to somewhere lower than $\frac{1}{2000}$ of the level which might occur without a filter.

It is important to incorporate response-tailoring circuitry in the early stages of an amplifier, otherwise large rumble signals may cause intermodulation distortion in later stages or in the recording process. To introduce filters after the damage is done is like shutting the stable door after the proverbial horse has bolted. A circuit for a microphone amplifier, based on a Mullard design but frequency-tailored to a response ideal for wildlife recording, is shown in Fig 37.

Incorrect polarity connections on microphones of the same make and type are not unknown. To ensure consistency in polarity, test the microphones to be used and if one is found to be phase-reversed have it modified: it requires only that the wires to the output pins be reconnected in the reverse order. In an organisation with a large range of equipment, phase problems are not uncommon. The main problem occurs with cables which have had an end connector pulled off and reconnected hurriedly with disregard for polarity and returned to service. The wildlife recordist should not be bothered by phase problems on cables because he will have lightweight cables on lightweight drums (or just coils) solely for his own use, and these he will guard zealously.

The cables are usually bound together in pairs for stereo working and in my experience suitable lengths are: one pair of 60m, a second pair of 40m and a third pair of 10m. It is advisable to mark the left and right cable of each pair, because marked cables help enormously when fault-tracing on a long cable run or when determining which microphone of a pair is not functioning properly. One good reason for using the microphones and cables the right way, in stereo terms, is that the aural information on the headphones supports the visual scene in the field. Knowing exactly what the sound is and where it comes from during recording will greatly assist in the processing of the recordings later on.

Phase checks should not be necessary in the field if care has previously been taken to check phasing on everything which is likely to be used. However, phase can be checked if the monitoring headphones or the recording machine has a means of combining the left and right channels. The Nagra 4S, for instance, has a push button to switch the left and right headphone feeds in parallel and phase checks can be carried out as follows: place the microphones closely together with their heads in line and record

191

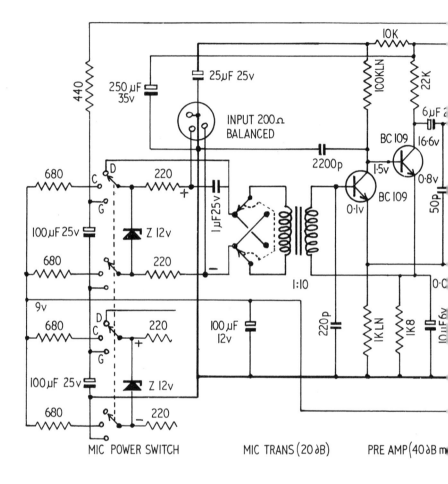

Fig 37 Circuit diagram for a high-gain microphone amplifier. It is based on a Mullard design but frequency-tailored to a response suitable for recording wildlife sounds (see text). The mic power switch is shown as a stereo coupled feed for: D, Dynamic; C, capacitor (9V phantom); G, Gun mics (12V line fed). Only one amplifier channel is shown. It draws only 2.1mA from the batteries

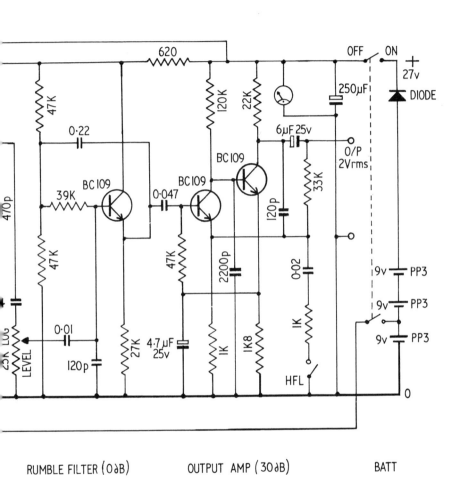

RUMBLE FILTER (0 ∂B)   OUTPUT AMP (30 ∂B)   BATT

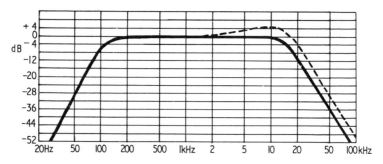

Fig 38 Frequency response of the amplifier in Fig 37. Broken line response is with high-frequency lift

a few words at about 1m distance. When phasing is positive little or no difference will be heard when the paralleling button is operated, but if the quality of sound becomes distorted or weak the phasing is negative and one of the microphones or cables is polarity-reversed. A short-term correction can be applied if the recording machine or microphone amplifier has a phase-reversal switch. Cables can be tested with a meter. Pin 1 at one end should go to pin 1 at the other end, and so on, right through every piece of cable likely to be used. Pay particular attention to interconnecting links between microphones and main cables.

With all the equipment tested and satisfactory, sling your recording machine on one shoulder and a coil of cable on the other, and off you go out into the wilds to enjoy the sounds of nature. Remember to take a spare set of batteries and plenty of tape—it could be your perfect day. When you return with masses of good recordings you will probably want to process them to master tapes—and I wish I could help you. Our next stage is master tapes and processing.

194

10

The Home Studio

The purpose of the home studio is to provide an environment suitable for the recordist to listen to and process his recordings. Many of the sounds recorded may have occurred either in non-reverberant locations or in surroundings exhibiting reflections peculiar to the area, eg the multiple echoes in a deeply wooded valley or the distinct echoes of sound across a large loch, canyon or gorge. To play sounds of this nature in a room which has its own strong reverberations will do nothing to enhance them, especially if the recordings are played in stereo or quad. It would confuse directional information and degrade sound quality.

Rooms which are dimensionally nearly cubical, with hard, flat walls and ceilings and uncarpeted concrete floors, are probably the worst, and speaking in them results in a very discomforting boomy sound. Smooth, flat ceilings, flat parallel walls and curved bay windows (which act in a similar way to parabolic reflectors) are all acoustically poor. Alcoves, recesses, non-parallel walls and sloping or stepped ceilings are more acceptable acoustically and form a good basis upon which to build a studio or listening-room. An acoustically good room need not be large either but the ratio of the basic dimensions is important. A satisfactory ratio is height 1 : width 1.6 : length 2.5. The minimum size for a good listening-room or small studio is about 4.5 × 3.5 × 2.5m high.

Reverberation time, the time taken for reflections within a room to decay to an inaudible level (or 60dB lower than the initial sound level), is determined by the rate of absorption of the sound by the surfaces in the room. When the absorption rate is low, sounds 'hang on' long after the original sound emission has ceased, causing the original and following sounds to be blurred in quality and directionally confused.

Acoustic treatment

The most obvious starting-point is the floor. A simple treatment is to cover the whole floor area in thick-pile carpet laid over thick underfelt. The ceiling might be treated by fixing proprietory acoustic ceiling panels over the entire area. Alternatively, the area could be treated with about 80mm thick glass-fibre insulation secured to the ceiling by some fire-resistant material such as chicken-wire. Chicken-wire and glass-fibre insulation are not visually very attractive so they may be covered with a more decorative material such as glass-fibre curtaining. This material could be stretched on to frames and the frames secured immediately below the chicken-wire and insulation. In a very high-ceilinged room a completely suspended decorative ceiling would probably improve the visual aspect. Panels of 2 × 1m would be easy to handle, simple to remove (for cleaning) and could be replaced with ease. These treatments are effective down to about 250Hz.

To absorb low frequencies thin panel resonators are required. The principle is that as panels vibrate under the influence of sound pressure waves they absorb energy. Once a panel is vibrating with the sound, the pressure waves meet no resistance and therefore are not reflected back as would be the case with a rigid panel. If after the sound ceases the panel continues to vibrate, it will then produce sound at its resonant frequency. To prevent this occurring, damping material is placed behind the panel. A panel vibrating in sympathy with sound-waves is the equivalent of an open window (no reflections). A panel resonator is not as efficient as an open window, but an efficiency of between 30 and 50 per cent can be expected. Sheets of thin plywood or fibreboard fixed to battens on the studio walls make good low-frequency absorbers. The panels should vary in size from about 0.5 × 0.5m to 2 × 1.5m to ensure an equal absorption over a wide range of frequencies. Recesses and alcoves are good places to install low-frequency absorbers. They may also be installed behind the decorative coverings of ceilings or fitted as an integral part of the final *décor*.

High/middle-frequency correction materials can be applied to walls in the same way as they are applied to ceilings and can be decorated to give a pleasing appearance. Alternatives are: perforated hardboard panels on battens with the cavity loose-filled with glass-fibre; carpet fixed over frames behind which there is carpet felt; one or two walls curtained in loose folds behind which are low-frequency absorbers and carpet felt. A good balance should be struck between the types of absorbers used because, ideally, the end result should be a studio with an equal degree of absorption throughout the audio frequency range 15Hz–20kHz.

A good listening-room should not be totally 'dead' but should feel comfortably non-reverberant. Windows are better double- or triple-glazed, with at least 80mm space between the panes to reduce outside sounds to a low level. They should also have full-length heavy curtains which can be drawn to reduce exterior sounds further and, at the same time, avoid sound reflections off the windows. Venetian blinds constructed from thin strips of metal are not recommended—they can resonate and so destroy the result of good acoustic treatment. Undamped springs on adjustable reading-lamps, wineglasses, thin metal panels and metal waste-bins can all resonate and cause interference. It will take skill and ingenuity to produce a studio with an ideal acoustic where sounds may be heard with precision and clarity and in which the listener feels comfortable and relaxed.

The equipment

In addition to providing facilities for listening to original and master recordings, the studio should be equipped to perform copying and frequency correction. To carry out either satisfactorily requires a good loudspeaker monitoring system. A high-quality loudspeaker with good low-frequency performance will show up many imperfections, like hum and rumble—imperfections which may pass undetected on a poor loudspeaker. A poor loudspeaker system in a good studio acoustic is as unfitting as a high-quality system in a poor acoustic.

For many recordists the studio may also serve the function of a living-room in which the recordist and other members of the household may wish to listen to a very wide range of sounds, including music, drama and possibly wildlife. 'Little boxes' may be satisfactory for listening to the final corrected masters of some wildlife sound recordings but, when such small units are used for assessing the original tapes or for listening to music, the only manifestation of the presence of large low-frequency signals is often the flapping of the cloth on the cabinet front.

A point to bear in mind is that loudspeakers which are capable of handling high powers are not necessarily capable also of delivering high-fidelity sound at normal room listening volume. Public-address loudspeakers can provide excellent clarity of speech at tremendous volume, but the sound quality of music played over the same system leaves a lot to be desired. Loudspeakers which sound impressive in the showroom may sound totally indifferent in the acoustically treated studio; therefore loudspeakers should, wherever possible, be obtained on approval and used in the studio for a week or more before deciding on their suitability.

A very good source of sound for assessing the merits of a loudspeaker is a VHF receiver, and human speech is probably the best material for forming a critical judgement. When testing a loudspeaker on speech the volume level should be adjusted to approximate that which would be obtained if those speaking were actually present. Male speech should at no time sound boomy or boxy at certain low frequencies—there should be no resonances—and breath noise and sibilance should sound natural. A system which reproduces sibilances harshly will quickly generate fatigue in the listener and cause him to reduce the listening volume or to apply a measure of treble attenuation to counteract the harshness. A stereo system switched to double mono (from both loudspeakers) should produce a needle-sharp central sound image.

The quality loudspeaker of yesteryear was a large and heavy affair. Then along came the great design revolution of miniaturisation and the market was flooded with very small boxes containing up to four miniature loudspeaker units—little boxes claimed to have amazing performances. Beside their big old-fashioned fathers, however, most of them failed dismally both in quality of sound and in sensitivity. More recently, loudspeaker design has shown a return to the larger sizes, incorporating most of the design principles of their forerunners and a few from their little brothers, and many of these larger units can produce good-quality sound. More attention has been given to improving the middle-frequency range and enclosures containing four or more separate units are quite common. However, this does not necessarily point to a four-unit loudspeaker being superior to a two-unit one at normal listening volumes. The multiple-unit systems may produce less distortion at high output powers; these may, therefore, be more favourably considered by a recordist who also intends to use his loudspeakers for a spot of disco work from time to time.

Where sounds of nature are his main interest, the recordist (now the listener) will find that realism is best conveyed when the reproduced volume level approximates to that experienced in the field—the natural level—and with certain subjects a volume level somewhat less. The correct listening level is most important when judging quality, balance and perspectives. They will all be more difficult to judge and control if the level is substantially higher or lower than the natural volume for a given sound and perspective. Local interference from noisy tape machines, grumbling heating pipes and buzzing clocks in the monitoring-room can affect judgement, especially where the acoustics are poor. 'Well, I didn't hear that in the field' are among the politer words uttered by recordists when they hear their recordings played on high-fidelity equipment in ideal acoustics.

Loudspeakers should be positioned with care. A stereo system which has one loudspeaker in a corner and the other standing free from acoustic obstruction may cause the balance of the bass to appear offset to the corner loudspeaker. If an unbalanced system is used to balance or control mastering, it is more than likely that the balance of the master will acquire an offset. The more advanced the recording system becomes, the greater the need to ensure that each channel in the system is closely matched. The closer the match the better the chances of producing a good master tape.

A few years ago vast numbers of mains mono tape recorders were to be found but now, due to the popularity of the compact cassette for domestic use, manufacturers have almost abandoned the cheap domestic open-reel tape machine. Fortunately though, many now produce the ideal machine for the enthusiast's home studio: the three-head stereo machine. Many three-head machines have built-in monitoring amplifiers of good quality and of adequate power (usually 10–20W) to drive the monitor loudspeakers in the home studio, where these are not separate. Tone controls on the replay (output) of a mastering machine are not necessary and can cause serious errors to occur when they are inadvertently left in circuit. Some form of quality control is, however, very desirable between the output of machines used for playing originals and the input of machines used for copying.

Response selection amplifiers (RSAs), are better unganged for stereo working because differences in quality, resulting from poorly-matched microphones for example, can be minimised by individual response tailoring. In wildlife recordings the calls of many species can be emphasised by the use of controls which 'lift' selected frequencies. Comprehensive response selection amplifiers provide both 'lift' and 'cut', usually at about 12dB/octave, at selected frequencies and the bandwidth of these sections is usually quite small. One type, known as a graphic equaliser, has a separate control for each of the ten or more selectable frequencies, which might be 40, 75, 100, 160, 240, 420 and 750Hz, and 1, 1.6, 2.4, 3.5, 4.8, 6.8, 10 and 15kHz. With a graphic equaliser it is possible to 'lift' a desired species while at the same time attenuating an undesired species or a noise in a different frequency band. Another type has a combined frequency selector switch and a separate lift/cut control; to perform a double correction operation it would be necessary to record two separate correction runs through the one unit or to use two units in series.

The normal treble and bass controls on pre-amplifiers and tape machines will not provide the sophisticated control so necessary to achieve good results from indifferent original material. For normal, straightforward music and speech recordings very little control should be necessary, except possibly to remove bass rumble, but for sound-effects and for natural sounds a very comprehensive control is desirable. Therefore, after a good-quality monitoring system and good copying facilities, a frequency response equaliser is the next most essential piece of equipment to have in the home studio. For many years I processed natural sounds with little more than bass and treble controls. Only during the last five years have really good control facilities become available. On many occasions, after recovering an obscure sound out of a seemingly incomprehensible mush, my regret has been that so much material was previously processed without such aid. Until a recordist has experimented fully with a really comprehensive equaliser on a whole range of material, the possible degree of refinement is quite beyond realisation.

The master machine, probably a three-head machine, should be correctly set for output levels, for frequency response and for azimuth. Avoid building up a collection of master tapes with recorded azimuth errors, for they can only cause problems later on when the tapes are played on correctly aligned machines.

Tone test tapes are available from many equipment and tape stores. They are rather costly but they provide a means of checking azimuth alignment of tape heads and frequency responses of replay channels. If the master tape machine is incorrectly aligned the quality of master/copy tapes recorded on it is automatically suspect.

Make sure that the master recorder does not introduce noise other than inherent tape hiss. If a bubbling noise is heard on the tape it is worth first changing the tape, demagnetising the tape heads, and then checking the bias settings on each track (bias, remember, is set to about 3 or 4dB over-drop at 10kHz, depending on the tape).

The home studio will obviously require a second tape machine but it need not be able to record: its function is to replay original material which can then be copied on to the master recorder. It is between these two machines that equalisation and level corrections are applied. A gramophone disc player might also be required from time to time, as would a cassette machine and a pair of headphones. Cassette machines can be considered standard home studio equipment, not so much for mastering but for recording and listening to cassette copies of masters. It is easier, less expensive and much safer to send cassette copies of masters to friends and other interested parties than it is to send the master tapes.

If the master tapes have for some reason to be passed out of your hands, high-quality safety copies should always be made.

It may be decided, for reasons of cost or space, to store all completed master tapes in cassette form—the master being erased and reused for mastering other material. In this case it would be wise to record at least three cassette copies of each master: one for your own listening, one for distribution among friends, and one to be held as an unplayed spare from which to make additional copies when your personal cassette gets mangled up. This stand-by spare and your own listening copy will, of course, be recorded noise-reduced in the interests of more pleasurable listening. The recordist may decide to master all his recordings at 19cm/s half-track stereo with noise reduction—this would indeed be a very high standard and one which is commercially acceptable.

Apart from the need for elaborate response tailoring in natural sounds and sound-effects recordings, a simple mixing system is sufficient for mixing the tape replay machine, the cassette machines and the disc player together to feed to the master recorder. Simple, three- and four-channel stereo mixers are available; they are not too costly and are quite suitable for home mixing applications.

Fig 39 shows the basic circuit of a four-input mixer. Only one channel is shown—for stereo it is necessary to double the circuit

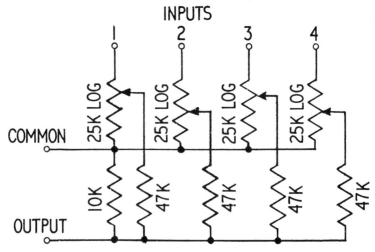

Fig 39 Circuit for a simple, passive four-input mixer

to two sections. If ganged, twin potentiometers are used they should be close-tolerance, otherwise the levels may vary and cause sound-stage positional shift as the control is operated. I used ordinary 20 per cent tolerance ganged volume controls but matched them to about 1 per cent tolerance. I explained my requirement in the radio shop and a whole boxful of ganged controls was placed at my disposal. It took about five minutes with a test meter to find four closely-matched controls.

The different types of level meters have already been discussed in an earlier chapter, but the point about levels and distortion applies equally in mastering. Many forms of material have transients which some meters cannot register and consequently distortion may arise on the tape before the meter shows much excitement; therefore on sharp transient sounds it is far better to be guided by the resulting sound quality than by metered levels. When copying difficult material it is necessary to do a test on the master to find the distortion point for that material and then to avoid reaching that point on the final run.

Processing to masters

A master tape is made up from material copied from one or more original recordings, with perhaps overlaid speech from a microphone in the studio. A master tape should preferably be free from any defects present on the original tapes—these having been edited prior to copying. A master tape may contain a few cut-edits but it is better to avoid them altogether.

Before any master copying is done it is advisable to play through all the original recordings to check their content, quality, perspective and balance, and also to time the sections considered for processing. A note should be made of defects and where they occur—a stopwatch is desirable for this operation. When all the originals have been listened to it should become clear what is and what is not usable, how it can be used and the order in which to use it. When all the originals have been edited and assembled in the chosen order each section should be checked for level settings and balance and, where equalisers are available, for quality adjustments. Where the original has a background relevant to the primary sound a sufficient background sound lead-in should be cut on to the beginning and end of each section, where it does not already exist, so that the introduction and exit of the section can be smoothly controlled. Sound effects, especially those of a surprise nature, should be leadered right up to the effect with marker tape so that not the slightest sound is heard before the effect.

A reference tone of between 400Hz and 2kHz at the front of a master tape will assist in setting levels for subsequent playbacks. The usual level on a stereo master is −12dB from peak modulation level on each track. The reference level should be noted on the box or in the documentation if the tape is destined for commercial enterprises.

When everything is ready to copy, the master recorder is run on 'record', in tape-out monitoring mode. At the appropriate moment the replay machine is started and, as the marked start of ambience or atmosphere on the reel passes the replay head, the mixer/fader control is advanced to the preselected setting, thus creating a fade-up introduction to the primary sounds. The length or rate of fade-up will depend on the nature of the material copied. The fade-up must not be on foreground subjects, except in drama scenes, but on available neutral atmosphere or background, so that the fade-in is completed before the primary material is reached. In processing wildlife sounds a reasonably slow and gentle fade-in and fade-out on atmospheres is pleasing. Birdsong should always start at the beginning of a song phrase.

In copying orchestral items there is little point in using the whole walk-on sequence of the leader and conductor with the audience applause unless, at the same time, there is a live speech introduction. Quite a pleasing effect is achieved by fading up the last five seconds of the applause followed by a short pause and music. The applause helps to set the scene a little better than fading up on two seconds of atmosphere before the music. The same fade technique can be applied to the applause at the end—take a few seconds of it and fade out slowly, reducing hf at the same time to improve the fade perspective. I use this hf-reducing technique in drama quite a lot because it produces a much more realistic distancing effect, particularly on musical effects and noisy battle scenes. Sometimes the right distancing effect can be achieved by reducing the amount of treble only, leaving the bass to continue at the original level.

At the end of a copy section a studio voice direct from a microphone or via tape may be required to assist continuity and to introduce the next recorded section. There should be no need to stop the master tape: the microphone or the recorded voice on another machine can be cued to start immediately the first one runs out, and so on in an alternating sequence between two replay machines or a replay machine and the studio microphone. It is possible to mix in subsidiary sources to mask poor jump-edits.

In the realm of wildlife recording it is possible to salvage a recording of a single species even if at first hearing it seems to be ruined by heavy background noise. Many birds have a very limited frequency range and by filtering it is possible to remove nearly all the obtruding background noise. However, with all possible background removed, the recording will sound somewhat lacking in natural atmosphere. This must be corrected by mixing with the salvaged recording a suitable neutral atmosphere which has itself first been corrected by removing some of the frequency band to be occupied by the subject—this is to allow the subject precedence. This sort of salvage operation can work quite well but it does require three machines—and the end result is nothing more than a fiddle!

It will be easier to spool quickly to a required section on a completed master if coloured leader tape is cut into the tape between each band of material. Where a large number of bands are involved different colours can be used to divide the tape into sections, themselves containing a number of bands.

Dub-editing

Dub-editing is electronic editing and master tapes so edited do not contain cut-edits. The desirability of a master recording machine capable of performing good dub-edits was discussed in Chapter 4; it was also mentioned that some machines produce a sharp click at the point of dub-over. Master tapes dub-edited on machines which produce clicks will obviously need to be cut-edited, finally, to remove the clicks. Dub-editing is less time-consuming than cut-editing because, when an edit proves unsatisfactory, no time is lost in running back a little and trying again. In cut-editing, a section of several minutes' duration may have to be re-recorded many times to achieve an acceptable edit.

A machine for dub-editing should have separate erase, record and replay heads: in other words, a three-head machine. Dub-editing is not editing in the proper sense because it cannot remove portions of modulation, as cut edits can: it is a system of building up material in a discontinuous fashion without recourse to physical cuts and joints.

A fault in an original recording can be eliminated by dub-editing. The tape is copied on to a master tape up to a point just beyond the fault, and the position of the fault is marked. The master is then run back a short distance before the fault mark. The original tape on the playback machine is marked at the point where the next copy-off section is to begin and it, also, is run back a little. Both machines are then run forward in the play mode, the replay output being heard from the dubbing machine. At the precise moment before the marked fault on the master tape passes the erase head, and just before the marked 'in' point on the original passes the replay head, the dubbing machine is switched to record mode—whereupon the result of the dub-on will be heard. If it happens to be unsuccessful this simple operation is repeated until a successful dub is achieved.

Some experimentation may be required to find the exact position for setting back the two tapes and the exact point at which to throw the dubbing machine to record mode, but once these points have been found further dubs can be performed with great accuracy and speed.

Dub-editing together two recordings with very different backgrounds will not produce an acceptable result, any more than would a cut edit; we have to smooth out changes in sound levels and sound quality. This requires two replay tape machines or a tape machine and a record player. If the outgoing sound track has a fairly long background tail and the incoming track has a long lead-in background preceding any primary modulation, a smooth mix between the outgoing track on one machine and the incoming track on the second machine is generally satisfactory. Where the outgoing and incoming tracks have only very short tails they can be spaced a little with blank tape and replayed on a single machine. A second machine (tape or disc) then plays a masking sound to hide the background jumps on the primary tracks. Just before the end of the outgoing track the masking sound is gently faded in, and as soon as the new incoming material becomes established the masking is faded gently away. The masking is assisted by quickly dipping the level of the outgoing track to reduce its background from a sudden cut to a steep roll-off. The front of the incoming track can be similarly rolled in.

It is possible to perform perfect dub-overs on two-head machines (normally unsuited to the purpose) by using a device as simple as a piece of cardboard. One piece of card is held between the tape and the erase head and another is held between the tape and the record/play head; then, by sliding the card out of the way at the appropriate point on the tape, the dub-over occurs. The machine has to be in the record mode for this operation and care must be taken to ensure that the card is thick enough to prevent erasure of previously recorded material. If the erase card is left in place and only the record card is slid out, sound-on-sound can be achieved. This method is very useful for superimposing additional material—voice-overs and special effects—because it reduces the level of high frequencies in the original sound (partial erasure caused by the hf bias applied to the recording head). However, the overlay sound is recorded with full response. The dub-over will not be heard simultaneously because only the incoming signal can be monitored and the tape has to be rewound to the dubbing point to check the result. Difficulties may be experienced in trying to card-dub on a cassette machine but it should be possible to card-out the erase head to do sound-on-sound.

Yet another dubbing possibility is the use of pause controls on the playing and recording machines. To be successful, both machines will need to have almost instantaneous forward run from pause. This method is reasonably successful on speech and there is normally only a small 'blop' on the master tape at the dubbing point, but where there is atmosphere of a fair level the result may not be altogether satisfactory. On atmosphere both the signal erasure point, caused by the record head, and the sudden cut on the outgoing material, caused by the tape remaining stationary at the erase head, can be heard. With a three-head machine the tape is not held stationary with the erase switched on. On dub-overs the erase and record currents start from zero and build up to normal in a few milliseconds; thus there is a smooth reduction of outgoing material and a smooth build-up of new material on the dubbed tape. This dubbing process may occupy as much as 5cm of tape at a tape speed of 38cm/s. When the erase head is carded out, sound-on-sound can be achieved. By experimenting with different card thicknesses it is possible to perform partial erasure of the existing material during the superimposition of the new.

Echo effects can be achieved with three-head machines during recording if a portion of the signal from the replay head is fed back to the input. A stereo machine can be used to generate interesting flutter echoes by delivering the output from one track back to the input of the other track, and using the original signal and the first and second delay signals.

Cut-editing

Physical cuts are made in recording tapes to remove unwanted sections of material from a recording and to build up a composite master tape. Cuts are made also to insert coloured leader tape. There are various methods of cutting and joining recording tape, but by far the simplest is that used by most professionals. For this a splicing block, single-edged razor blade and a roll of adhesive splicing tape are required. Tape-splicing blocks are precision machined to grip the recording tape pushed into them and this enables cuts and joins to be executed with speed and ease.

When a fault or the decided cutting point on the tape is reached, the playing machine is stopped and the tape is manoeuvred backwards and forwards (not too slowly) by hand-turning the spools—the tape running against the replay head. The material on the tape will then be heard and the precise location of a fault or the decided cutting point can be found and marked. The return point is marked similarly. The tape is then carefully pulled away from the heads and at one of the marked points it is pushed in the splicing block and cut. The unwanted section is then pulled off until the other marked point is found, and cut. The two ends of tape will then lie in line in the splicing block and a small length of splicing tape is applied to complete the edit. The join is carefully pulled out of the splicing block and replayed.

A very sharp cutting blade should be used to avoid tape creasing—the thinner the tape the more likely it is to crease. Cutting blades should be free from magnetism: cutting with magnetised blades causes bumps on replay.

A splicing block usually has 90, 60 and 45° cutting guides. Which to use will depend on the nature of the recorded material and on the tape speed. A 45° cut may prove unsatisfactory on speech editing at 9.5cm/s in stereo, often resulting in positional shifting, whereas a 60° cut may be satisfactory. On some atmospheres, however, a 45° cut may turn out to be better than a 60° cut, because a 45° cut produces a slower rate of change at the edit point and is thus smoother and less noticeable. When edits prove unsatisfactory, it is often worthwhile trying an alternative cutting angle. 90° cuts are rarely used, except for joining on metallised tapes for switching functions, because they produce square-edge modulation bumps on replay; the nearer vertical a cut the greater the bump. If both tracks on a stereo recording have to be cut simultaneously, two 45° lead-in half-cuts should be made in such a way that one side of the edit has a 90° arrowhead and the other side a 90° vee about the centre line of the tape. The modulation then enters each stereo track at the same instant but avoids a square-edge bump.

To allow joints to flex easily across the heads the edit joints should show a hairline gap when held up to the light. A very tightly butted or slightly overlapped joint will result in a drop-out, owing to the tape lifting off the head at that point.

Splicing tapes of the correct width for reel-to-reel tapes and for cassette tapes can be bought at most tape shops. There are two adhesive types: one for smooth-back tapes and another for matt-back tapes. Only 2 or 3cm of splicing tape are necessary to make a joint. It is carefully aligned over the top of the two pieces of tape (which are held in perfect alignment in the splicing block); the free end is allowed to drop on to the tape in the block groove and, when it is perfectly in the centre, thumb or finger pressure is applied and followed along the block groove, making positive adhesion. The splicing tape should be slightly narrower than the recording tape and it must not be allowed to fall off-centre and stick out over the tape edge, because it may adhere to the next layer of tape on the spool and cause bounces as it releases with a snatch. Some splicing tapes ooze adhesive after a while under pressure. To prevent snatches and tape bounce (and on machines with pressure pads to prevent the sticky edges of the joints from pulling on the pads), all the edit joints on a recording may require dusting with French chalk.

There are a number of different ways in which to mark recording tape, depending on its colour and type. Matt-back tape can be marked with felt-tipped pens, ordinary lead pencils and hard crayons. Smooth-back tape is more difficult to mark and usually requires a soft wax pencil. Soft wax is easily transferred to tape heads, tape guides, capstan and pressure pads (where these are used). Wax on any of these can cause tape binding, which often produces a nasty rasping noise and in extreme cases can even stop tape motion completely. It is important, therefore, to mark the tape only lightly to prevent build-up of wax deposits on the guides, and frequently to clean all component parts on the tape path with solvents.

Black marking-pencil wax (used by some tape editors on light-coloured matt-back tapes) is very difficult to see on black ferrite heads. When taking over a studio previously used for editing I have had to hold up many a recording session in order to clean tape heads thoroughly. All heads and guides may appear perfectly clean, whereas, actually, a very thin film of wax deposit is covering them and causes hf loss. A soft cloth and solvent rubbed carefully over the heads and guides provides visual proof of the presence of a substantial coating of wax. It only goes to show that master recording machines should not be used for tape editing—it wears the heads unnecessarily and marker-editing can clog the heads.

I was in an audio equipment showroom recently when a customer complained that the cassette machine he had bought the previous week was giving a poor performance. The assistant, probably being used to complaints of this sort, proceeded to apply a small quantity of solvent to a cleaning swab and thoroughly cleaned up the machine. The performance was then up to the customer's satisfaction and he went away with a smile.

The assistant showed me the cleaning swab and it looked as though it had been used to mop up cocoa. That, of course, was not wax but shed tape oxide—a further reason for frequent cleaning.

It is not essential to mark a tape in any way and many professionals dispense with marking altogether, preferring to pull the tape away from the replay head at the exact point at which a cut is to be made, using a thumbnail as a guide. Offset marking may be used if the heads are not accessible. Some machines have a small marking-post installed somewhere to the take-up reel side of the heads and on machines additionally provided with an editing block a corresponding offset mark is inscribed on the block. Even with offset points it is not necessary to use marking pencils—a thumbnail will do just as well and is much quicker.

The practice of marking tape with wax pencils, especially the marking of lines of considerable length along the tape while it is playing, should be resisted. Excess wax on the back of tape sooner or later results in small quantities being transferred to the oxide side and then to the tape heads and guides.

Making unnoticeable edits on atmosphere and background recordings can be difficult: the first attempt often results in a slight kick or jump in the sound and additional cuts may be necessary, removing more material, until the edit proves acceptable. It is better to cut immediately before and really tight up to the beginning of a foreground sound than after one. The new incoming sound directs a listener's attention from the background and thus the primary sound masks the change (if any) in background. If the incoming material is a fairly prolonged loud sound of wide frequency content, substantial changes in background may be quite acceptable, for masking relies totally upon the frequencies of the foreground and background sounds. A high-frequency foreground sound will not, on an edit, mask the introduction of a low-frequency background sound, and vice-versa.

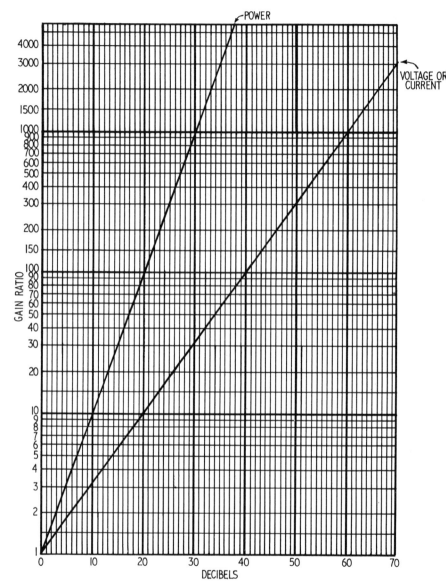

Fig 40 Conversion graph showing power and voltage ratios with their equivalents in decibels

Glossary of Terms

Acoustic table — Table with mesh or perforated top which is acoustically transparent

Ambient sounds — Surrounding sounds

Attenuate — To make less

Axis — Imaginary line of symmetry (in the case of microphones, the front—the line of maximum sensitivity)

Azimuth — Vertical alignment (of magnetic gap in a tape head)

Back-to-back — Back-to-back cardioids—two cardioids facing in opposite directions

Baffle — A separating screen

Balance — A state of equilibrium. Equal volume levels. In stereo, left/right equilibrium

Blasting — Noise caused by a blast of air on a microphone

Capstan — Rotating spindle which drives tape

Carding-out — Placing a piece of card between erase head, recording head and tape to prevent erasure of existing signals

Coincident — Identical (or near-identical) positions of two or more microphones. Not spaced

Coloration (of sound) — Acoustic or electronic reverberation (echoes), causing deterioration in clarity and intelligibility of sounds

213

dB (decibels)   A convenient logarithmic measure for sound levels and voltage ratios

Dead zone   Angle of minimum sensitivity

Directivity   Directional selectivity

Drop-out   Tape drop-out—momentary loss of signal due to tape oxide irregularities

Equaliser   Frequency response-tailoring unit

Equaliser, graphic   Where the frequency response is displayed visually by the physical positions of the many equaliser controls

Erase head   On tape recorder, for removing existing signals prior to recording new

Fade-board   A piece of board used mainly in stage drama work to attenuate volume levels from artistes and sound effects—a large baffle

Flutter   Quick variations in speed or level

Flutter echo   Quick repetition

Focal point   The point of concentrated signals

Fundamental frequency   First-order frequency—the basic frequency, not an overtone

Harmonics   Overtones of the fundamental

Howl-round   Fierce high-volume feedback between loudspeaker and microphone, producing a loud howl

Impedance   Resistance—in inductive and capacitive circuits, resistance varies with frequency

Leader tape   Coloured tape for inserting into recordings to separate and identify different bands or items

Masking   The dominance of one sound over another

Microsecond (μs)   One-millionth of a second

214

Millisecond (ms)	One-thousandth of a second
Mixer, active	Has separate amplifier for each source, therefore no reaction occurs between different inputs. The better mixers have balanced transformer inputs
Mixer, passive	Interconnected electrical resistance network, with no individual input isolating amplifiers
Muffs	A microphone muff is a small, external windshield to reduce blasting and mic pops, caused by breath blasts
Multi-tracking	Building up multiple-sound recordings in easy stages by playing existing sounds, as a guide, while recording additional sounds
Notching	Hearing defect in which an individual suffers a reduced sensitivity to certain frequencies of sound
Octave	A frequency ratio of 1 : 2
Oxide	The material on the recording side of recording tape
Perspective	The sound balance ratio of a near source and a more distant one
Phantom-powered	Power to microphones, etc, can be supplied via the cable, either along the signal pair or between the signal pair and the screen of the cable
Phase polarity	Positive phase: polarities the same. Negative phase: polarities different
Polarity	Positive or negative direction—ways of connecting a pair of signal conductors: in phase or phase-reversed
Polar response	The response of a microphone to sounds from different directions
Pressure pads	Spring-loaded pads of soft felt which

press the tape against the heads to ensure intimate tape-to-head contact on a recording machine

Quad	Quadraphonic sound—four-channel surround sound
Radio microphone	A microphone with an integral radio transmitter (or these can be quite separate items), the only wire used being a few centimetres of aerial
Reverberation	In a room: multiple echo reflections. Electronically: generation of ambience echo
rms	Root-mean-square—the root of the mean of the squares of the current/voltage values of a waveform. It is the alternating current equivalent of the direct current which would produce the same heating effect in a given resistance
Scale distortion	Frequency distortion when sounds are listened to at volume levels much greater or much less than their natural level
Sound-stage	The imaginary stage area between the loudspeakers in a stereo listening system
SPL	Sound pressure level
Splicing block (editing block)	A block of metal machined with a wide groove in which to place recording tape for cutting
Splicing tape (editing tape)	Adhesive tape used to join the cut ends of recording tape
Stereo	Two separate but interrelated signals emanating from two loudspeakers equidistant from, and presenting a reasonably wide angle to, a listener
Superimpose	Lay one sound over another

216

Test tape

A high-precision tape of frequency tones for the alignment of tape head azimuth and for checking replay and recording responses

Transient

Very short duration. A good transient response handles very short-duration, fast-rising signal levels. A poor transient response is a slow-to-act response and routed signals sound dull and lifeless

Voice-over

Speech superimposed over other sounds

Wow

Slow-rate fluctuations of speed, pitch or level

Index

219

221